THE HISTORY OF CIVILIZATION

THE ECONOMIC LIFE OF THE ANCIENT WORLD

THE HISTORY OF CIVILIZATION

General Editor C. K. Ogden

The *History of Civilization* is a landmark in early twentieth Century publishing. The aim of the general editor, C. K. Ogden, was to "summarise in one comprehensive synthesis the most recent findings and theories of historians, anthropologists, archaeologists, sociologists and all conscientious students of civilization." The *History,* which includes titles in the French series *L'Evolution de l'Humanité*, was published at a formative time in the development of the social sciences, and during a period of significant historical discoveries.

A list of the titles in the series can be found at the end of this book.

THE ECONOMIC LIFE OF THE ANCIENT WORLD

J. Toutain

Translated by
M.R. Dobie

London and New York

First published in 1930 by Kegan Paul, Trench, Trubner
Reprinted in 1996, 1998, 2002 by Routledge

11 New Fetter Lane
London EC4P 4EE
&
29 West 35th Street
New York, NY 10001

Routledge is an imprint of the Taylor & Francis Group

Printed in Great Britain by
Antony Rowe Limited, Chippenham, Wiltshire

British Cataloguing in Publication Data

ISBN: 0-415-15585-1
ISBN Roman Civilization (6 volume set): 0-415-15613-0
ISBN History of Civilization (50 volume set): 0-415-14380-2

TO MY DEAR WIFE

PAULINE TOUTAIN-BLANCHET

IN MEMORY OF OUR FATHER

D. BLANCHET

CONTENTS

PART II

THE ECONOMIC LIFE OF THE HELLENISTIC WORLD

PART III

THE ECONOMIC LIFE OF THE WESTERN MEDITERRANEAN AND THE NEIGHBOURING REGIONS TO THE ESTABLISHMENT OF THE ROMAN EMPIRE UNDER AUGUSTUS

PART IV

THE ECONOMIC LIFE OF THE ANCIENT WORLD UNDER THE ROMAN EMPIRE

MAPS

FOREWORD

ECONOMIC DEVELOPMENT: METHODS AND FACTORS

Here is a volume of great importance in our general plan. In it we reach back to the volume on Prehistoric Man, *in which we saw the part played by the* hand, *the continuations of the hand, the making of the first tools, the birth of industry and its expansion, the appearance of agriculture, and the beginnings of trade and its development by the mutual relations of human groups.*

As we said in the Foreword to that volume, our lamented friend Paul Lacombe was to have written an introduction establishing the connexion between these two works; he was to have dwelt on the evolution of technique, on the part played by economics in the history of mankind. At one time Paul Lacombe, who afterwards admitted the superior importance of the intellectual element, was inclined to place economic activity in the forefront of history. As far as the beginnings were concerned, he was absolutely right. Without ceasing, " the air devours man "; he has to repair himself, to remake himself. Sun, rain, and wind compel him to cover himself with a roof, with a garment. " Only when the work of economics has been done do the other desires unfold, and, according to the amount of effort which has been required for economics to reach their end, there is more or less time and will left for all the rest."[1] *His reflections, such as the doctrine of " historical materialism,"*[2] *had the good effect of causing historians to devote much attention to an order of phenomena which had hitherto been too much neglected. The history of* Homo Faber *had to be written, and that history was all the more interesting and necessary in view of the fact that, as our friend said, " man has at all times been to a great extent a workman."*[3] *We have promised to dwell especially, taking our in-*

[1] *De l'histoire considérée comme science*, p. 48.
[2] See our *Synthèse en histoire*, p. 181.
[3] *Prehistoric Man*, pp. xiii-xiv. " So the hand, which met the immediate and imperative needs of the body, decided that man should be above all things a manual worker, a tool-maker, a digger of the earth.

spiration from him, on the part played by work, on the succession of inventions, " on the infinite development of tools, born of the hand, the consequences of which are infinite in themselves, often quite unforeseeable ; tools have made man like a god."

After Prehistoric Man, *in which M. de Morgan, with his immense knowledge, embraced primitive civilization in all its fumblings and all its extension, a certain number of volumes—those of MM. Febvre and Pittard, which dealt with physical environment and racial migrations, and those of MM. Moret and Davy, Moret, Delaporte, and Glotz, which brought Egypt, Mesopotamia, and the Ægean on to the historical scene*[1]*—have contributed to revealing the part played in the remote past by the need to keep alive, and then by the desire to better material life. Here, in a specially economic study, in a huge general picture, M. Toutain, who knows the Græco-Roman world thoroughly, shows how classical antiquity, taking over the achievements of other times and places, strove to turn the terrestrial environment to better and better account—down to the days of disaster, ruin, and a temporary retrogression of collective life.*

To write the economic history of societies is, indeed, as has been seen and as will be seen more fully here, to follow the work done by man to improve his dwelling (οἶκος, οἰκουμένη), *to make use of his domain, to profit by its various resources—objects and their properties, superficial or deeper, and energies of all kinds—to increase the number of " goods," of " values," and to form plentiful reserves of them ; and it is, at the same time, to study the distribution of those goods among men.*

Many problems arise from a study of this kind—the part played in economic development by accident in its various forms ; also, and chiefly, the part played by the social factor, the logical factor, individual inventiveness ; the methods of organization governing the production and circulation of goods or their distribution ; the effects of luxury on civilization and the relationship of economic progress to human happiness. In economics,

---

And that is what he still is, for far the most part. That, it seems, ought not to appear in the history of man, and that is just what is real history, truly universal or semi-universal history—but it is the history known as historical materialism " (*Journal*, 26th July, 1915).

[1] *A Geographical Introduction to History, Race and History, From Tribe to Empire. The Nile, Mesopotamia, The Ægean Civilization*, all in this series.

as in other things, comparison, based on the facts recorded by the historian, entails generalization.

M. Toutain, like some others of our collaborators, would write purely as a historian. In the setting of the Mediterranean basin and the outlying regions which were attached to it by the Roman Empire in its greatest days, from the beginnings of Greek history to the fall of the Empire, he describes "economic life in all its forms, in its almost infinite variety and complexity"; he sets forth "facts, hard, definite facts, with reference to their place and date as far as possible."[1] *He is afraid of over-simplification;*[2] *he is mistrustful of comparisons;*[3] *he refuses to systematize, for to attempt to extract systematic views from the manifold, diverse data is to leave the solid ground of "true history."*[4]

But the mere extent of the subject gives rise to comparisons and, by the nature of things, general ideas force themselves upon the author or upon the reader. They are all the more valuable for not being preconceived. The philosophy of history, in its generalizations, is inclined to be impatient; historical synthesis, as we have preached often enough and have endeavoured to illustrate in our practice, proceeds cautiously and safely, from the particular to the general.

While, then, this book is a remarkably accurate historical treatise, which, by the host of details which it gives about material life, enables one to conjure up the daily existence of the ancients in all classes of society, we shall here dwell not so much on that aspect of the book as on the results which emerge from it; we shall systematize, in spite of M. Toutain—and with his help.

First of all, what our fellow-worker's general plan and the manner in which he has carried it out show brilliantly, is the close connexion between economic life and social morphology, or the changes which take place in the size and density of societies.

Often, in previous Forewords, we have called attention to the fundamental identity, the innate sympathy, the disposition to help one another, which tend to bring human beings into association, and also to the unifying process, largely due to war

[1] Below, pp. 1, 4. [2] P. 29. [3] P. 26.

[4] Without, of course, refusing to make use of the works of other modern writers, he prefers, with the school of Fustel de Coulanges, to resort as much as possible to the ancient authors and documents.

and conquest, which makes historical groups, from the clan to the empire. The advance of association and unification had its effect on economic life, but it was itself to a great extent brought about by economic interest. The desire to have more room and the love of adventure, in individuals and in groups, are not in themselves sufficient to explain migrations, nor can ambition alone explain conquests. Men join men and groups join groups, or subdue others to themselves, in order to improve their material life. As, more or less consciously, a society grows bigger and denser, its solidarity becomes tighter and more complex, and differences become more marked and produce unforeseen effects.[1] *A whole economic life, what may be called an economic complex, can only be studied and defined in relation to social structures, which are themselves bound up with a logic and also with a host of contingent circumstances. This emerges clearly from M. Toutain's account.*

At the beginning, the Greeks lived a mainly rural life, " bounded by a very close horizon," " tied to their own soil." They were dependent on the Phœnicians, " at once merchants and pirates, sea-traffickers who had taken the place of the Ægeans in the Eastern Mediterranean."[2] *The Greek colonization which went on from about the eleventh century to the end of the seventh was the great fact, due to many causes and productive of many consequences, which utterly transformed the economic life of the Mediterranean. The Greeks extended the area from which they themselves obtained resources—foodstuffs and raw materials—and opened markets for their own products, but they also took to sea-faring and enterprise, winning " the mastery of the waves and sea-routes "*[3] *and extending the area of exchanges from one end of the Mediterranean to the other. " Before that great event, the expedition of Alexander, before all barriers between Greece and the East were thrown down, the enterprising genius of the Greeks had penetrated west and north far beyond the points where it had succeeded in founding colonies and trading-stations."*[4] *Work became more intense; industry and, above all, trade developed; hence the new part played by money, that admirable*

[1] See Durkheim, *De la division du travail social*, esp. pp. 336 *ff.*; Lacombe, *op. cit.*, p. 192; Cornejo, *Sociologie générale*, pp. 468 *ff.* (see especially interesting remarks on what he calls " moral density "); *La Synthèse en histoire*, pp. 135 *ff.*, 184 *ff.*

[2] Pp. 20. 23. [3] P. 30. [4] P. 70.

instrument of exchange, and the constitution of movable wealth by the side of landed wealth, which, however, still held first place. Athens now played a preponderant part, and in general the cities, which continually increased in number, had a great influence on the economic organization of the world.

The extension of the area of exchanges, which was at once a cause and a result of the increase of consumption, went on under Alexander and in the Hellenistic period. We have seen the marvellous plan of the unity of mankind conceived by the pupil of Aristotle.[1] *That unity was destined to be only a dream, but the result of his expedition was a very real coming together of the peoples. Greece gradually decayed, but the wealthy East, exploited and rendered fruitful by the Greek genius, with towns springing up and rapidly rising to prosperity, great trade-routes developing, and centres of importation, exportation, and transit appearing on every side, attracted the products of distant and sometimes hardly explored lands and reached the markets of Central Asia, India, Arabia, the east coast of Africa, and Western Europe. Kings and cities seem to have become aware of " an economic solidarity previously unknown."*[2] *Into that prosperous world flowed all the things which could satisfy men's needs—or provoke them. In illuminating pages we shall see what delights Hellenistic civilization already offered to the privileged—and what temptations to the rest.*[3]

Almost to the beginning of the Christian era, the Western Mediterranean presents a very different picture. When the East was in full bloom, the West was still going through the first stages " towards the rational use of natural resources, towards the methodical and fruitful organization of human labour."[4] *In Carthage, which played the same part in the West as Tyre in the East, perhaps in a still more narrowly and sordidly mercantile fashion, and in Italy, which was well situated, economically and racially, to serve as a connecting link between West and East, progress was made, the chief factor in which was the sea.*[5] *Rome " was to preside for several hundreds of years over the economic synthesis and expansion of the ancient world,"*[6] *and the foreign policy of Rome was to make economic life, " if one may use the words regarding antiquity, international and world-wide."*[7] *The*

[1] *Macedonian Imperialism*, in this series. [2] Pp. 139-41, 152, 161.
[3] Pp. 157 *ff*. [4] P. 204. [5] Pp 199, 212.
[6] P. 213, *cf*. p. 251. [7] P. 226.

foundation of the Empire had the same results as Alexander's conquest. It widened "the horizon of the activity and labour of man to the very boundaries of what the ancients called the Inhabited Earth, the Οἰκουμένη*"; only the Atlantic was an "impassable barrier."*[1] *As M. Toutain justly observes, from the Christian era onwards, progress was no longer bound up with conquest, but with the Roman Peace whose character and blessings are described in M. Chapot's* The Roman World *and with the work of internal organization—improvement of the administration, diffusion of law, multiplication of cities, and development of the system of routes by land and sea.*[2] *There were two centuries of remarkable prosperity during which the transformed West and the East became more and more united. Comfort—"but it was still a very comparative comfort"—extended to the mass of men, while luxuries were obtained from various sources, far beyond the Mediterranean basin.*[3] *In the second century, under the Antonines, this economic evolution touched its highest point.*

If we mentally change the position of the chapters which M. Toutain devotes to the Western Mediterranean (in which he gives a full, straightforward summary of the progress accomplished in the stone and metal ages by certain peoples of Europe), we shall obtain from this volume, in an abstract scheme, a succession of main stages passed in economic evolution—from nomadic life *(in which man goes to the place of production) to* settled life *(in which he attracts products to the centres of consumption)*[4] *and from* household economy *(production and consumption in common) to* urban economy *(division of labour among producers and consumers, town markets) and then* inter-urban *and* international economy *(monetary economy and exchanges over greater and greater distances). Moreover, once the Greek world has reached a certain degree of economic organization and expansion—down to the advent of machinery and the creation of true capitalism—the characteristic features of economic life are not again altered in their fundamental nature; but there were serious crises, followed by centuries of retrogression, the many causes of which will be described in the volume on* The End of the Ancient World and the Dawn of the Middle Ages.[5]

Here we are doing no more—as may be seen—than calling

1 Pp. 250, 259. 2 Pp. 259, 284, 303, 323.
3 Pp. 291-3, 297-8, 307. 4 Pp. 176-7. 5 Pp. 79, 323 *ff*.

attention to the results of a piece of historical work which preaches no doctrine. With François Simiand we shall readily say : " Historical reality is complex, and cannot be brought under formulas. There are differences between similar phenomena found in different societies and at different times. I do not deny all this, but the fact does not invalidate the legitimacy of comparative studies or of the establishment of types, if these are undertaken with a suitably critical spirit." With him, again, we shall say : " It is not going away from reality and plunging into logical, artificial constructions, to analyze that reality with clear but exact categories and to look in it for relations which are general in form but are always close to the data supplied by fact ; for only these categories and relations give a true understanding of the very complexity of that reality."[1] *Now, history likewise compels M. Toutain, in regard to the classification of economic phenomena and the evolution of phenomena of various categories, to make interesting statements which we have now to underline.*

In each division of this book, the chapters agriculture, industry, trade *recur regularly. They correspond to the various activities by which goods come into being and move about. The two great classes of economic phenomena, production (agriculture and industry) and exchange or circulation, are discussed more fully than distribution ; phenomena of this last class, though not neglected, are perhaps given less importance, and in any case are not studied by themselves.*[2]

[1] In *Année sociologique*, vol. x, pp. 533, 551, regarding works by Salvioli and Mantoux.

[2] We are almost in agreement with F. Simiand over the classification which he has slowly elaborated in the first series of *L'Année sociologique*, and described in greater detail in *La Méthode positive en science économique*, pp. 153 *ff.*, and *L'Année sociologique*, new series, vol. i, pp. 720 *ff.* We believe, however, that it is better, at least in the case of ancient economic life, not to include *circulation* under the head of *production*; but we agree with him that *distribution* and *consumption* tend to merge into one another. To study consumption is to study the use of wealth, immediate use, answering more or less well to the need (lack, over-production), and deferred use (conservative or productive). To study distribution is to consider the users. Here we cannot follow the classification of Simiand in its details, which are complicated and perhaps, in some points, provisional.

The problem of classification interests many economists, who are trying to break down the traditional framework and to conform to the nature of things. For this question and for economic bibliography in general, use *L'Année sociologique* (economic sociology), the *Revue de l'Institut de Sociologie, Solvay* (political and social economy), and the

From M. Toutain's book it emerges plainly that in ancient times productive activity, considered in its two kinds and in respect both of technical methods and the details of institutions, stopped or slowed down in its advance sooner than commercial activity.

Agriculture and stock-breeding, with the beginnings of settled life, with the clearing of waste land and deforestation, were at first the chief labour of man. Little by little, the two kinds of farming developed, as the Greek world and then the Roman Empire extended, bringing wider and wider areas into use, and as methods were improved, the rational utilization of the soil and climate making it possible to increase the yield.[1] *The work of man—small landowner, agricultural labourer, or serf attached to the soil (as a result of invasions)*—created *wealth which became more and more plentiful; for nature has no* value *except for the need and by the effort or idea which utilizes and exploits it. But a time came when that exploitation stood still. There was even a shortage of corn at the end of the Empire. The* latifundia, *which owed their development to slave labour, after having been the curse of Italy, became the curse of the Roman world. The indifference and greed of the great landlords led them to neglect corn, and the system of* colonatus *was devised in order to meet the danger of famine. As for agricultural apparatus, it underwent little change, and "the real implements of the farmer still had their old shape. Plough, spade, hoe, mattock, pick, fork, scythe, sickle, and pruning-knife were, as the surviving specimens show, just as they had been handed down from generation to generation."*[2]

Industry, "that is, the transformation of various raw materials into things intended to satisfy men's needs and luxurious tastes,"[3] *progressed—still more than agriculture—by the specialization of the crafts and a division of labour which went to great lengths.*[4] *It is in the domain of industry that that important social phenomenon manifests itself most effectively. We see the* workman

handbook entitled *Die Wirtschaftstheorie der Gegenwart: I. Gesamtbild der Forschung in den einzelnen Ländern: Frankreich*, a remarkable general survey of French works and tendencies, taken in conjunction with many foreign works and tendencies, by G. Pirou.

[1] Pp. 87, 102, 106, 109, 260, 265-6, 271. In the words of a Greek philosopher, intelligence became "the fellow-worker of nature." See Cloché, *La Civilisation athénienne*, p. 135.

[2] Pp. 262-3, 282.

[3] P. 17.

[4] Pp. 47-8, 53-6

appear, " whose labour, being devoted to one definite task, supplied the other members of the community with the food, utensils, tools, and manufactured articles which came to be needed all the more as the production of them became more normal, more plentiful, and more technical. So the energies of each man were released from the manifold tasks for which they were not fitted, and could concentrate and so become more fruitful."[1] *Family industry was superseded by small and medium industry. Workshops and building-yards were set up, and even fairly large factories, under the Empire.*[2] *In Rome, professional corporations came into being and, after first, it seems, being confined to handicraftsmen, extended to manufacturers and merchants. " The idea of association was one of the original features of the economic life of Rome."*[3]

It may well be imagined that the number of raw materials, of things utilized, steadily increased. The urban centres of the Greek world and the Empire consumed an infinite variety of substances for building and furniture-making.[4] *Yet this industrial development had its limits. There could not be a true " big industry," for lack of machinery.*[5] *In sum, industrial methods did not make any decisive advance in the centuries covered by this volume, any more than agricultural methods. Such progress as there was consisted chiefly " in the extension of industry, in the number of workshops built and in the* number and extension of the markets *opened to it."*[6]

We must lay stress on this last point. Down to the decline of the Empire, trade did not cease, not only to extend, but to perfect its methods. Freedom and ease of trade grew steadily as well as its volume.[7]

No doubt, if one considers the evolution of trade from its beginnings, there does seem to have been a stage of social constraint at which it was subject to all kinds of religious restrictions and regulations ;[8] *but it developed in a non-religious, individualistic direction,*[9] *and above all in the direction of international organization. It has even been maintained that trade must have*

[1] P. 57. See Cloché, *op. cit.*, p. 137. [2] Pp. 301 *ff.*
[3] P. 316. [4] Pp. 130, 284, 286-7, 292.
[5] P. 57. [6] P. 293. [7] Pp. 29, 62, 139-40, 195.
[8] See Davy, in *From Tribe to Empire*, pp. 108-109.
[9] Huvelin, " Définition et évolution du droit commercial," in *Rev. de Synthèse historique*, vol. vii, p. 72.

been international (" *intersocial* " *would be a better word*) *before it was internal, sea-borne before it was land-borne; but the theory seems to hold good only for certain complex forms of business. However that may be, at an early date, without doubt, division of labour and mutual exchange of services tended to become established between men belonging to different and even hostile groups. We know of the phenomenon of the* " *silent trade* " (*goods laid down on a beach*), *which for the time being makes the foreigner* (hostis) *a party in an exchange, an associate.*[1] *Periodical meetings, or markets, strengthened by bonds of hospitality, give exchange a regular character. The system of* " *community of market* " *must necessarily end in* " *more or less complete fusion of the groups resorting to the market. The market provided with common organs constitutes the centre of crystallization of a new society based . . . on differences and division of labour. Round the market, merchants, craftsmen, etc., establish themselves. The peace* (*the law*) *of the market extends to them permanently, and you have* a city. *The city is a market whose establishments have become fixed and whose peace has become permanent. The agglomeration of an* urban *form is characteristic of a commercial civilization. The market of the city* (agora, forum, *etc.*) *for a long time continues to be the centre of common life.*"[2] *Between groups which become larger and larger and more and more distinct, trade continues its work, the importance of which cannot be exaggerated. It establishes a solidarity among men, even over great distances.*[3] *The men who did trade on a big scale, the traders of Phœnicia, Greece, Syria, Carthage, and Rome, not only assumed a capital part in the evolution of all economic life, by causing every form of production to increase; they contributed to the unification of the ancient world, and by causing tastes, customs, and ideas to circulate, as it were, they served the unity of mankind.*

What was at once the cause and the effect of that splendid development of trade, was the continuous creation in the Mediterranean basin of means of expansion, of the material equipment needed—routes by land and water (*sea, rivers, canals*), *vehicles*

[1] *Ibid.*, p. 67.

[2] *Ibid.*, pp. 68, 74. For the Middle Ages, see G. Bourgin, " Les Origines urbaines du Moyen Âge," in *Rev. de Synth. hist.*, vol. vii, p. 321.

[3] Below, p. 140

and ships, various driving-powers, skilfully constructed ports, and huge entrepôts.[1] *At the same time, commercial institutions, especially in the Roman Empire, "attained a high degree of perfection."*[2] *Compared with urban trade, the early booths and shops, and even the "market-halls, local markets, and fairs,"*[3] *the trading and shipping offices which were built up on the corporations of merchants in the great commercial centres represented a very great power. "From one end of the Empire to the other, commerce was, as it were, unified. Natural commodities, raw materials, and manufactured goods travelled about the Mediterranean in every direction; they were carried by whole merchant navies, and associations of shipowners . . . doubtless played the part of the modern shipping company."*[4] *Certain pages in this book, at the end of Parts Two and Four, give a very definite idea and sometimes a striking picture of the organization effected under Alexander's successors and the Roman Emperors and of the prodigious activity and variety of the circulation of goods.*[5]

But the essential agent of this intensive trade, and therefore of increasing production, was money. Even when the exchange of gifts and barter properly so called were superseded by the use of metal, raw or manufactured,[6] *a real progress was accomplished. But "the really fruitful revolution, which gave trade an impulse which it had not known before, was the invention of true money, coin with a public, official stamp."*[7] *There was now a common measure of the* value *of things,*[8] *which was convenient because of its portable, easily handled form and also because of the impersonal character which it gave to dealings:*

[1] Pp. 147, 241-4, 312.

[2] Huvelin, "Histoire externe et interne du droit commercial," in *Rev. de Synth. hist.*, vol. viii, p. 201.

[3] See pp. 66 and 158 for remarks on the roads used by traders and pilgrims, which should be compared with J. Bédier's ingenious suggestions regarding the function of the pilgrim-ways in the Middle Ages.

[4] P. 319.

[5] Pp. 158-9, 306-7, 317 *ff.*

[6] Pp. 22, 70-1, 186.

[7] P. 71.

[8] Subjectively the value corresponds to the need, the desire. In addition to this subjective element, there are, in the notion of value, objective elements—plenty or scarcity and work. Hence the opposition between demand and supply. The measuring of values, in economy, is a remarkable and paradoxical invention; economic value, as Simiand has ingeniously said, is *an opinion which is a quantity*. See *La Synthèse en histoire*, p. 186, and *Année sociologique*, vol. xi, p. 543.

" it made them an operation between two abstract entities, the buyer and the seller, who did not need to know one another."[1] *Subsequent progress consisted in a tendency to unify currencies (if monetary unity was first instituted by Rome, the preponderance of Athens had made the Attic drachma a kind of standard),*[2] *but chiefly in the various uses of coin which led up to the banking system, and then in the organization of banking and the development of the mechanism of finance—public and private banks, current accounts, cheques, letters of credit, transfers from one account to another, and joint-stock financial companies.*[3] *" By the side of that first instrument of circulation, money, another appeared, namely* credit. *It facilitated circulation because it made it possible to exchange, not only present wealth against present wealth, but present wealth against future wealth."*[4]

So movable wealth was formed, and so " capitalism " came into being, if it was not the capitalistic system as we think of it today.[5]

Really, it is a mistake to confine the word " capital," as is usually done, to movable wealth. " Capital " includes everything which, resulting from previous work, is not consumed immediately, everything among the natural objects used by man which is to serve for later satisfactions. Land under culture and working equipment, like coin, where everything is potential, eminently represent the fruit of human experience and foresight. Thanks to capital, man lives less and less at the mercy of chances, coincidences, circumstances ; he makes certain not only of well-being but of unlimited possibilities of better-being. All men, however, do not share equally in the immediate consumption which most meets the need ; still less, then, do they share equally in capital which accumulates. The ownership *of things is distributed unequally. And the unequal distribution of goods is a most important object of study in economics.*

1 Cornejo, *op. cit.*, vol. i, p. 481.

2 Pp. 73, 167. Cæsar created the gold standard, the *aureus.*

3 Pp. 74-6, 165-8, 246-50.

4 Huvelin, *art. cit.*, in *Rev. de Synth. hist.*, vol. vii, p. 69; below, p. 79.

5 The capitalistic system is the application of accumulated capital to industry, with the effects on the organization of labour which it involves. In speaking of antiquity, one can at the very most talk of commercial capitalism, and one can just talk of financial capitalism. See H. Sée, *Les Origines du capitalisme moderne*, p. 7.

At the beginning, was ownership collective? Lacombe has observed that the scholars who devote their studies to the classical peoples " seem to doubt that their *peoples ever practised collective ownership." Sociologists, on the contrary, " assert that all peoples without exception began with collective ownership, and that even the peoples in whom the scholars in question specialize still practised it at the dawn of the historical period, or, at the very least, still possessed institutions which were clearly descended from that practice. I take my stand firmly," Lacombe declares, " on the side of the sociologists."*[1] *His argument is very strong; it refers, however, only to the soil. For the times and peoples with which he deals, M. Toutain is disinclined to admit the contention of the sociologists, or, at least, he holds that individual or family ownership existed alongside of the other kind. However that may have been, we here follow the development of private property and see the rise of all the inequalities in natural resources, and therefore in pleasures and leisure, which gave rise, in the course of history, to countless economic conflicts and moral problems.*

It would seem that in Greece, where land continued to be the main form of wealth until a late date, a certain equality was maintained fairly long. Before the fourth century, according to Aristotle, nobody was in utter distress.[2] *In the Hellenistic period, landed property became concentrated. In Rome, at an early date, the distribution of land was very unequal; there were rich and poor, and no middle class, but a town proletariate. The countryside, for causes which are indicated in this book and set forth more fully elsewhere,*[3] *was deserted. The " agrarian question " became a grievous problem; the laws passed to remedy the evil " were openly violated or quietly evaded."*[4] *In the Empire as a whole, small and medium sized property held their own for a long time. But when other countries were afflicted by the same curse as Italy, when the small landowner and free tenant farmer disappeared, as in Egypt, where the system of farming-out had been established for the royal and sacred domains, the State had to interfere to prevent the land from being deserted, and created the new class of* coloni, *slaves of the land.*

[1] *L'Appropriation privée du sol,* in *Rev. de Synth. hist.,* vol. vi, p. 24.

[2] P. 39.

[3] See *Primitive Italy,* in this series.

[4] P. 221.

Movable wealth was as unequally distributed as the soil. If we consider industry, except in the East, where there were many free craftsmen combined in guilds (but subject to strict regulations and fairly heavy taxes), and in some Greek cities,[1] *it did not serve to spread ease of life or to diffuse the fruits of capital, for slaves played a large and disastrous part in it, both driving out the free worker and taking the place of machinery. It was one of the essential characteristics of ancient economic life that there were human beings who were treated as things, that the manufacturer had at his disposal (like the big landlord, of course) a capital of flesh and a working plant of muscle. On the " unprecedented extension " of slavery, as a result of wars and piracy, and on the low cost of slave labour (hard enough in other respects), M. Toutain gives edifying details.*[2]

Antiquity, as we know, was unacquainted with machinery. Decisive progress had been made in technical methods in the stone and metal ages.[3] *Why, between those distant times, in which a series of marvellous inventions provided economic life with its essential stock-in-trade, and the age of machines, do we find increasing multiplication and circulation of consumable things without any parallel advance of the " manufacturing intelligence "?*[4] *Are we to suppose that the aptitude declined because humanity proceeds by alternate periods of technical invention and speculative crisis? We have already discussed this* law of the two states, *which is very contestable and too absolute.*[5] *No doubt, social organization can, at certain times, by the*

[1] Pp. 60-1.

[2] See esp. pp. 231-2, 289. *Cf. Bulletin de la Soc. franç. de Philosophie*, Feb.-March, 1914, " Y-a-t'il un rhythme dans le progrès intellectuel ?" (Meyerson), p. 103: " A strong slave . . . was worth on an average, in the Roman world, £16; but there were periods . . . when his price went as low . . . as 3s. Cato the Elder allows " (his slaves) " no rest; when a slave is not eating or sleeping (and of course he should do as little of either as possible), he works." On the condition of the slave and his gradual betterment, and on the freedman, see Declareuil, *Rome the Law-giver*, in this series, pp. 126-35, 340-42.

The craftsman seems to have complained, not of the competition of the slave, but rather of the need for industrial labour—on account of the lack of land.

[3] P. 174. *Cf.* Weber, in a/m *Bulletin*, pp. 81-82; Lacombe, *op. cit.*, pp. 162, 173.

[4] We must not deny all invention, all progress in details; see, in a/m *Bulletin*, Meyerson, pp. 95 *ff.*, and Weber, p. 118. *Cf.* G. Lombroso-Ferrero, " Pourquoi le machinisme ne fut pas adopté dans l'antiquité," in *Rev. du Mois*, May-June, 1920, p. 451.

[5] Foreword to *Prehistoric Man*.

progress of "verbal technique," by "illusory techniques," of religious, magical origin, and by the conservative power of tradition and the corporate spirit, hamper the play of the "mechanical instinct" which is born in the individual at the contact of nature and has benefited social organization. But it seems to be above all slavery that is to blame here. Not only did it supply a lazy solution for technical problems (Aristotle said that the slave would be indispensable so long as the shuttle did not fly by itself);[1] *it caused manual labour to be despised as a servile occupation. By a kind of contamination, the material applications of science seemed degrading to the Greek and Roman philosophers. Archimedes himself, according to Plutarch, regarded "mechanics in general and everything practical as a low and obscure art."*[2] *Although many guilds of small manufacturers and free craftsmen sprang up under the Empire, workshops of any size continued to rely on slave employees to supply the needs (small as they still were) of the masses and the luxuries of the few. Moreover, the object of the professional associations was chiefly moral. In short, in the presence of slavery, there was not in antiquity, and could not be, any machinery or any organized body of wage-earners.*

If, now, we turn to trade, Rome is different from Greece and the Hellenistic world. In the latter, a large class of merchants profited by its ample development; in the former, it was left to a minority of big business men, who did not enjoy great esteem. It was held in the same contempt as the mechanical arts, and down to 224 B.C. *it was, like them, forbidden to all Roman citizens. The Senatorial order continued to be officially excluded from it (Claudian Law). Outside the great companies run by members of the Equestrian order, "the greater part of business . . . was in the hands of freedmen, usually of Greek or Oriental origin."*[3] *They flooded Italy with Eastern goods, but with far more luxuries than articles of general consumption. In the face of the Senatorial class which had grown rich by conquest and pillage, they constituted a plutocracy, equally powerful and pleasure-loving.*

Wealth in Rome, which originally came from the right of the stronger, then from big business and banking, but chiefly from

[1] Meyerson, *art. cit.*, p. 104.
[2] P. 239. *Cf.* Meyerson, *art. cit.*, p. 101; G. Lombroso-Ferrero, *art. cit.*, p. 459.
[3] P. 245.

the farming of taxes[1] *and usurious money-lending (the legal interest was* 12 *per cent.), had too little to do with agricultural and industrial work. In general—and one should especially note this—antiquity was incapable of making really fruitful and beneficial use of movable capital. No doubt, the formation of cities, states, and empires, the increase of the size and density of human societies, and the impulse of social logic which tends to intermingle, to fuse particular interests in ever greater numbers, the better to satisfy them, led, in the ancient world, to the development of production and circulation, to the creation of the mechanism of money and banking, to more well-being, and to a future secured for a longer time ahead. But technical processes had remained stationary, or almost so. The capital amassed by human labour and its fruits were distributed haphazard. Events showed how fragile was that economic constitution, how precarious that prosperity—and how curiously unequal, too, in different times and places. Before the economic crisis and retrogression, a moral crisis had set in, just because of the contrast which there was between excess of selfish wealth and extreme penury ; beyond the problem of wealth, of earthly felicity, it raised the question of human happiness.*

It is well known that economists for a long time erred in the direction of pure theory, deducing the laws of economic development from the definition of Homo Œconomicus. *Interest in history and the progress of ethnology and sociology have to a great extent added to and corrected the hard-and-fast orthodox conceptions. But in the place of over-simple deduction new theories have risen up. Misusing ethnology, some have made a too systematic reconstruction of economic beginnings from the so-called " primitive man " of today.*[2] *Misusing sociology, others have explained economic evolution too exclusively by the social factor.*[3] *The complex interplay of the various factors of that evolution—manifold accidents, institutional necessities, social and intellectual logic—is clearly seen in this very objective*

[1] Until the system of direct collection was introduced and wealthy taxpayers were made responsible (the result of which in the end was that they migrated into the country).

[2] Just remarks in O. Leroy, *Essai d'introduction critique à l'étude de l'économie primitive* (on K. Bücher).

[3] See *La Synthèse en histoire*, pp. 181 *ff.* (on Durkheim).

study, which surveys economic reality widely and from a great height.[1] *M. Toutain has contributed to our synthesis a stone which is all the more valuable in that it falls into place—let us say it once more in concluding—without any effort to propound a doctrine.*

HENRI BERR.

[1] A curious and significant feature—M. Toutain believes in the effective part played by the individual (see pp. 328-9), and rightly. We have tried to show (in *La Synthèse en histoire* and General Introduction to *The Earth before History*) what the individual can do, either as a contingent or as a logical factor in the social and in the intellectual order. But in this book, except perhaps Alexander, we hardly *see* one individual, so overwhelming are the great economic phenomena, the mass phenomena. *Cf.* the remarks of Henri Sée, in a similar work, *Les Origines du capitalisme moderne,* in which he has "always endeavoured to rely on concrete facts," to take accidents into account, but has endeavoured "to do a work of synthesis"; he observes that "personalities fall entirely into the background," in favour of "certain facts of another order," whose "connexion" is more apparent (pp. 2-3).

THE ECONOMIC LIFE OF THE ANCIENT WORLD

INTRODUCTION

DEFINITION OF ECONOMIC LIFE. ITS HISTORICAL CHARACTER. THE ECONOMIC LIFE OF THE ANCIENT WORLD: GREAT PERIODS, DOCUMENTS, METHOD

TO study the economic history of a given period is to inquire what were, during that period, the organization of property, agriculture, and industry, the development and main directions of routes and means of transport, the procedure and character of exchanges, and the amount of consumption—in short, to describe economic life in all its forms, in its almost infinite variety and complexity.

In all countries and at all times, economic life has been strictly subject to physical, racial, and social conditions. Also, in any one people, it develops in the course of the ages. It is therefore necessary, when one starts to study that of a particular portion of mankind, to determine the exact setting of the subject in place and time.

In the ordinary language of European science, ancient life is that which developed chiefly round the Mediterranean basin. Even the civilizations of the ancient East came into some kind of contact with the sea which the Romans afterwards called, with justice, *mare nostrum*—Our Sea. It is true that there came a time when that life, with its special characteristics, spread over countries whose rivers flowed into other seas, whose climate was unlike that of the Mediterranean lands—Lusitania, Gaul, Britain, and the Rhine valley in the West ; the deserts of Syria and Arabia, the banks of the Euphrates, and the shores of the Indian Ocean in the

East. But those regions turned round, as it were, to face the Mediterranean, which attracted their economic activity and held it in its own orbit for many centuries. The unity of the Mediterranean, with these annexes on either side of it, was accomplished by the Roman Empire, the only state which has succeeded in bringing together under one authority such distant and unlike countries as Britain and Egypt, Dacia and Mauretania, Gaul and Arabia.

Before arriving at that unity and concentration, Mediterranean life developed in a continuous process which is marked by the progressive reduction of local autonomies and differences. That development did not fail to have a profound influence on economic life. After the works of Messrs. de Morgan, Moret, Delaporte, and Glotz, on *Prehistoric Man*, *The Nile and Egyptian Civilization*, *Mesopotamia*, and *The Ægean Civilization*, our starting-point can only be the beginning of Greek history properly so called. The first organized society which presents itself to our view and our investigations is that depicted in the Homeric and Hesiodic poems. At that time, the Mediterranean was still divided into compartments, no doubt not separated by watertight barriers, but each keeping its own physiognomy—the Eastern Mediterranean, bordered by the native civilizations of Egypt, Phœnicia and Syria, and Asia Minor; the Greek Mediterranean, confined to the coasts of the southern parts of the Balkan Peninsula and the islands of the Ægean Sea; the Italic and Sicel Mediterranean, the inhabitants of which had not emerged from the stage of prehistory or protohistory; and the Western Mediterranean, whose seaboard was divided between Phœnicians and Carthaginians, Iberians and Ligurians, and Etruscans. As the ages went by, two powers, two civilizations, gradually extended their domain, Hellenism in the East and Rome in the West, until the day when Rome, victorious in war and policy but conquered morally by Greece and, later, by the East, created the unity of the vast sea-basin at the centre of which she stood. That unity lasted four hundred years. When it collapsed under the repeated blows of the barbarians from the long unknown depths of Northern and Eastern Europe, ancient times were at an end.

This outline of general history allows us to define the

great periods into which the economic history of antiquity should be divided.

The first period is the Greek period, properly so called, during which Hellenism, while extending its sphere of action from the north of the Euxine to Cyrenaica and from Rhodes and Cyprus to Marseilles and Saguntum, nevertheless remained strictly national in character.

The second period, which opens with Alexander's expedition into the East, is the Hellenistic period. Checked in the West by the advance of Rome, the expansion of Greece found the field free in Asia and in the Nile valley all the way to the Indian Ocean. New routes by land and sea were opened to economic activity. The centre of gravity of the Greek world shifted eastwards and southwards, from Athens, Corinth, and Sparta to Rhodes, Antioch, and Alexandria.

While this centre of intense life was developing in the Eastern Mediterranean, in Italy Rome was growing greater, subjecting all the peoples of the peninsula to her rule, one after the other, triumphing over Carthage, and gathering under her sway all the live forces of the West. The growth of the Roman Republic did not take place without grievous crises. It was accompanied by an evolution which was often disturbed, retarded, or hampered by violent episodes, and the origin of evolution and crises alike must be sought in the economic conditions of public and private life.

Hitherto East and West had followed each its own destiny. No doubt, they were not unaware of each other's existence. They did not stand back to back; commercial relations had already started between the various coasts of the Mediterranean. But the political unity of the Mediterranean world had not yet been effected. This was done in the first century before Christ. It became more and more complete and stable as Rome gradually set up her administration, her Proconsuls, Legates, Procurators, and Prefects, in the place of the old local and regional sovereignties, whether royal lines, federations of cities, or free towns. Henceforward the economic field of action was one single domain, the *orbis Romanus*, in which all peoples worked and prospered side by side—Orientals, Greeks, Italians, and the barbarous or semi-barbarous peoples of North Africa, the Iberian Peninsula, Gaul, Britain, and the valleys of the Rhine and Danube.

Over every frontier, the Roman world entered into relations with distant markets. There were barriers which could not yet be crossed, the Atlantic in the West, the Rhine and Danube, and the plains of southern Russia; but the Caucasus and Iran in Asia and the sands of the Sahara in Africa were passed, and the Indian Ocean bore mariners to India. This last period, which covers the first four centuries of the Christian era, really closes the economic history of the ancient world. From the fifth century onwards, the successive shocks of Germanic, and then Arab, invasions broke up the Roman unity, subjected economic life to new conditions, destroyed its continuity, and directed it towards quite new destinies.

I shall, therefore, divide this study into four main parts:

I. The economic life of Greece and Greek lands, to the expedition of Alexander.

II. The economic life of the Hellenistic world, to the conquest of the East by the Romans.

III. The economic life of the West, chiefly in Italy and Rome, to the end of the Roman Republic.

IV. The economic life of the Roman Empire, to the barbarian invasions.

In this geographical and chronological setting, what I propose to set forth is, above all, facts, hard, definite facts, with reference to their place and date as far as possible. In order to obtain and establish these facts, we have at our disposal an abundance of evidence of all kinds—works of ancient authors, literary, historical, geographical, and technical, legal texts, inscriptions, papyri, coins, and archæological remains. We shall find plenty of information in poems like the *Iliad, Odyssey, Works and Days*, and *Georgics*, in romances like Longinus's famous pastoral *Daphnis and Chloë*, in histories like those of Herodotos, Thucydides, Polybios, and Livy, in geographical works like those of Strabo, Ptolemy, and the *Geographi Græci Minores*, in technical treatises like some of Xenophon's minor works and those of Cato, Varro, and Columella, and in encyclopædias like that of Pliny the Elder. The Roman Codes and the inscriptions will tell us about the organization of landed property and industrial labour. The papyri are full of facts about the daily life of Ptolemaic and Roman Egypt. Coins will be our witnesses for trade. What we can still see of the quarries, mines, metal-works, potteries, roads, and harbours of Greek and Roman antiquity will be of the greatest help to us in reconstructing the outward appearance and material equipment of economic life.

To these documents we shall apply a strictly objective method. We shall study them as they have come down to us, without trying to complete them by restorations which are often rash or by hypotheses which may well obscure or distort the true meaning of texts. We shall have the wisdom to stop in the presence of a gap in our information, and sometimes to say, "We do not know." In my opinion, this is the only method which will enable us to give solid foundations to the modest construction which we are attempting to set up.

For, from the facts which I shall have succeeded in assembling, it is my intention to extract general ideas. But I shall be careful not to present in that guise any abstract theories, evolved *a priori*, which would have no value but that of personal opinions formed irrespective of historical reality. My only ambition is to draw up, so far as it is possible today, a balance-sheet of the economic life of antiquity. This ambition may appear too lofty to some, and too modest to others; but it is in pursuit of it that I would make use both of my own labours and of the teaching and counsels which I have received from my predecessors and masters.

PART I

THE ECONOMIC LIFE OF GREECE AND GREEK LANDS TO THE EXPEDITION OF ALEXANDER

CHAPTER I

HOMERIC AND HESIODIC SOCIETY

WHATEVER may be the dates of the *Iliad*, the *Odyssey*, and the Hesiodic poems, these works all depict one same society, whose economic organization is marked by special features of its own. It is not easy to indicate the beginning of the period in which this society lived or to determine the circumstances in which it succeeded the society of the Ægean age, but it is at least certain that it is earlier than either the settlement of the Dorians in the Peloponnese or the expansion of the Greeks over the Mediterranean. Achæans still occupy the valleys of the Eurotas, Pamisos, and Alpheios, the mountains of Arcadia, the plains of Elis, and the peninsulas of Argolis. Crete, Rhodes, and some islands of the southern Ægean—Nisyros, Carpathos, Cos—have come into the Hellenic domain,[1] but the peoples dwelling in Thrace and on the western coasts of Asia Minor are included in the *Iliad* among the allies of the Trojans.[2] In the *Odyssey*, Alcinoos tells Odysseus that Eubœa is the furthest of the lands known to the Phæacians.[3] It is true that Hesiod, in his *Works and Days*, relates that his father, after vainly seeking his fortune at Cyme in Æolis, took ship and settled in the small Bœotian town of Ascra.[4] Even if we admit that the so-called Æolian colonization had by that time reached the coast of Asia Minor, we know that, according to Greek tradition, that settlement was earlier than the great movement of the Ionians.[5] As for the countries of the West, Trinacria, that is Sicily,

[1] *Il.*, ii, 645 *ff.*
[2] *Ibid.*, 816 *ff.*
[3] vi, 321 *ff.*
[4] 625 *ff.*
[5] **XXX**, i, p. 145.

was still unknown to Greek mariners, and they regarded Italy as the abode of divine beings like Circe. Homeric and Hesiodic society, then, had for its geographical setting Greece Proper, and for its chronological setting the period extending from the end of the Ægean age to the great migrations by land and sea which gave the Hellenic world its final shape, constitution, and extent.

I

Agriculture and Fruit-Growing

In the economic life of this society, the growing of the fruits of the soil held a very important place. Every great estate includes cornfields and vineyards. Corn and vines are frequently mentioned as the chief kind of agricultural wealth. The corn grown is wheat, barley, and millet. No banquet is complete without wine, and no religious ceremony without a libation of wine. In his description of the shield of Achilles, the poet of the *Iliad* represents the harvest and the vintage, as typical of the work of the fields.

> "And he made thereon a rich domain" (of a king). "There were reapers mowing, with sharp sickles in their hands, and along the swathes the sheaves fell thickly to the ground. . . .
>
> "And he made a vineyard, full of grapes, fair and golden. There were black bunches hanging, and it was set all over with vine-poles. . . . And merry lads and maidens bore the honey-sweet fruit in plaited baskets."[1]

On leaving Ithaca for Pylos and Sparta, Telemachos takes as provisions for the voyage twelve jars of wine and twenty measures of the purest flour in well-sewn skins.[2] On the shield of Heracles, as on that of Achilles, harvesters and wine-gatherers personify rural life.[3] In the *Works and Days*, the growing of corn and wine takes the first place.

Orchards and gardens are rich in fruit and vegetables. The olive, pear, apple, fig, and orange are the trees mentioned most often. They abound on the estates of Alcinoos and Odysseus.

[1] xviii, 550 *ff*. [2] *Od*., ii, 353 *ff*.
[3] Hesiod, *Shield of Heracles*, 286 *ff*.

II

Stock-Breeding

With agriculture, stock-breeding is one of the chief sources of wealth. Bulls, oxen, cows, and heifers, rams, ewes, and lambs, goats and kids, horses, mares, foals, and mules, hogs and sows, fill stables and byres or gambol in meadows, at the bottom of fertile valleys, on the slopes of untilled hills, and even in the undergrowth of forests. Some furnish man with their flesh for food and their hides or wool for clothing; others give him their milk as well; others lend him their strength, to draw chariots and carry loads. Swarms of bees produce honey and wax for him. Of poultry, only geese are mentioned.

III

Forests. Hunting and Fishing

Between or around the fields and pastures stretch vast forests. They contain oaks, pines, firs, poplars, alders, and beeches, and sometimes also laurels, cedars, and cypresses. Along the watercourses, willows and osier-beds appear.

There game is plentiful, and hunting is one of the chief pleasures of men. They hunt deer (stags, roes, fawns), wild-goat, boar, and hare. They also come upon wolves and bears, and even lions. In the air, they try to hit thrushes and doves.

Fishing is done with the net or with the hook.

IV

Farming Methods and Equipment

The technical methods of agriculture and stock-breeding were already fairly developed, and an equipment existed, still very simple, but adequate.

To meet the exhaustion of the soil, it was allowed to lie fallow every other year, each piece of land being divided into two breaks, one of which rested while the other was sown.[1]

Corn-growing required two ploughings, and perhaps three, the third being done after the sowing, to bury the seed.

[1] XLV, p. 471.

The furrows were cut straight, with a plough made all of wood, usually drawn by two oxen, or sometimes two mules, which the ploughman urged on with a goad. After the ploughing which followed the sowing, the ground was gone over with the mattock, to cover any seed which the plough might have left on the surface. The bigger clods were broken up with a beetle.[1]

The harvesters used the sickle. When mown, the ears were collected in sheaves, and these were carried away on carts, probably like that described by Hesiod, a sort of waggon consisting of wide, low body on two wheels. The corn was threshed on a round threshing-floor, well exposed to the wind. Oxen and mules went round the floor, trampling on the sheaves. The straw was lifted frequently, that no ear might escape the treading of the beasts. Beating was also done with the flail. When the grain was collected it was put in jars, and then shut up in the barn.[2] Flour was ground either with a pestle and mortar or with a quern, and was stored in skin bags.

Vine-growers sometimes—if not always—trained the branches on props, so that the grapes did not lie on the ground, but hung from the shoots at some height. The chief operations of vine-growing were the pruning, the second dressing, and the vintage. The grapes were gathered in baskets, and then left exposed to the air for fifteen days, first ten days in the sun and then five in the shade. Then came the treading and the vatting. From the vat the wine was poured into jars, where it was kept.[3]

Fruit-growing and vegetable-gardening were practised in an equally intelligent fashion. In the orchards, the soil at the foot of the trees was turned up. Irrigation provided the gardens with the water which they needed. The use of manure seems to have been known.

Livestock was raised chiefly on natural prairies and uncultivated land, and beasts were also led into the forests. It is not impossible that artificial meadows existed, kept green and fertile by irrigation. These doubtless provided fodder and bedding, which were stored in the barns as soon as the harvest was in.[4] During the good weather the flocks and

[1] *Ibid.*, pp. 475 *ff.*
[2] *Ibid.*, p. 480.
[3] *Ibid.*, pp. 481, 485 *ff.*
[4] Hesiod, *Works and Days*, 514.

herds remained out-of-doors, but when winter came they were taken to the byres, which were an indispensable appendage of the farms in a society in which livestock was one of the chief forms of wealth. On the estate of Odysseus, there were twelve pigsties round the house of Eumæos.[1] The farms possessed dairies for milk and cheese. The author of the *Odyssey* stocks the cave of Polyphemos with all the material needed—pots, pails, tubs of whey, and crates laden with cheeses.[2]

The axe, the saw, and the wedge served for dealing with the trees of the forest. The hunter's weapons were the boar-spear and the long spear. Game, furred or feathered, was also taken in the net, and hounds were used for starting hares and attacking boars and other savage beasts.

Who supplied the labour for all these operations ? There is no doubt that the owners of estates, large and small, worked their land themselves when necessary, or were at least able to guide a plough and handle spade, sickle, and scythe. Old Laërtes works in his garden and tends the trees of his orchard. Odysseus challenges the suitor Eurymachos as follows:

> " If we should vie, which of us could do the more work in the grass, in the spring season, when the days are long, I should have my well-curved sickle and you would have the like, and we should mow without eating till the dusk, so long as there was any grass. And if we had oxen to drive, big, fair, and well-fed, equal in age and strength and of the same size, and a field of the same size and kind to plough, you would see whether I could drive a straight furrow !"[3]

When Odysseus utters this challenge he is, no doubt, disguised as a beggar; but would he utter it if he were not really able to stand the test as a reaper and as a ploughman ? In any case, Eurymachos, to whom the challenge is addressed, is a landowner. In the *Works and Days*, the poet is speaking to the owner of a small farm, who toils in person on his fields, among his vines, and in his barn and byre.

But the work of the master alone was not sufficient for the exploitation of the soil. Many men were needed on the big estates. Some of these, at least, were slaves, men and women (δμῶες). But there were also agricultural labourers of free condition, *thetes*, who were taken on for special work—harvest, vintage, olive-picking—or for stated periods, short

[1] *Od.*, xiv, 13 *ff.* [2] ix, 218 *ff.* [3] *Od.*, xviii, 366 *ff.*

or long. Hesiod advises his brother Perses to engage a *thes* to look after his barn when the harvest has been ingathered.[1]

As a rule, the beasts were entrusted to the keeping of slaves. Might one not, however, infer from the story of Polyphemos that the owner of a herd or flock sometimes led it to the pasture himself? True, it may be objected that Polyphemos is not a typical member of Greek society, but surely the poet has invested the imaginary world of the Cyclopes with habits taken from the society in which he himself lived.

V

The Organization and Character of Landed Property

What was the true character of rural landed property, and how was it organized? The answer to these two questions would be very simple if it was enough to look for it in the Homeric poems and Hesiod. But—and here lies the great disadvantage of rash hypotheses—we have first to clear the ground of certain theories which complicate the problem which we have to settle. On the ground that "collective ownership of the soil is the natural form of ownership among primitive peoples,"[2] many scholars, especially jurists, have endeavoured to prove that the Homeric Greeks, while acquainted with individual ownership of land, also practised the collective ownership of arable land, and even that this latter was the normal, common form of ownership among them. It appears to me that this reasoning contains a serious error of method and a no less serious error of fact.

First, for the error of method. Because in some primitive peoples ownership of the soil has been collective, that is no reason why the same system should have existed uniformly in all primitive peoples. Those who make this conclusion forget that the character of land-ownership cannot be independent of the nature of the soil, nor of the climate. It is stated, for example, that the organization of agrarian property in the Russia of the nineteenth century can and should help us to determine the system upon which such property was

[1] *Works and Days*, 594. [2] A. Esmein, in **XVI**, 1890, p. 822.

organized among the Greeks of Homeric and Hesiodic times. Or, again, we are confronted, for the same purpose, with the evidence which Tacitus gives us about ownership among the Germans. This is making very light of the essential differences which exist between regions like the vast plains of Russia and Northern Germany and the peninsulas and islands of Greece. There you have boundless open spaces, here, small compartments; there, an unending plain without any natural limit, here, territories bounded by mountains or the sea; there, a continental climate of hard winters and sometimes torrid summers, here, a maritime climate of temperate heat and cold. It seems to me impossible that the system of ownership should not reflect these contrasts in physical conditions. In any case, a method which would, in such a matter, draw conclusions from one country to another is in my opinion thoroughly dangerous. As Monsieur P. Guiraud has said with his usual vigour and clarity: " It is useless to say that this " (the system of collective ownership among primitive peoples) " is a general law of mankind. A law is true only if it agrees with the facts, and with the facts the last word must lie."[1]

Then there is the error of fact. The Greek people, as painted in the Homeric and Hesiodic poems, was not a primitive people. We cannot accept the opinion which Esmein expresses in these terms: " The customs to be observed in the *Iliad* and in the *Odyssey* have the clear-cut character of primitive customs."[2] The discoveries which have been made, since the day when those lines were written, in Argolis, in Crete, in Bœotia, in the Troad, and in the islands of the Ægean have proved that Homeric society was anything but a primitive society. " From the mixture of Cretan and Hellenic elements the Mycenæan civilization was formed. It carried on the Cretan civilization, but it must have been less brilliant, because the whole of Greece was still upset by the invasion and by incessant wars."[3] Whatever the Achæans may have done and suffered in the course of their taking possession of the land of Greece, neither in manners nor in organization, social, economic, or political, do they look in the least like primitive men.

[1] **XLV**, pp. 1 *ff*. [2] Esmein, *loc. cit.*
[3] **XLVII**, English p. 74.

The result of this twofold error is that, to justify it, the texts have been simply tortured. M. Guiraud has no difficulty in demonstrating this in the first chapters of his *Propriété foncière en Grèce.* After setting forth and refuting the arguments used to establish the existence of agrarian communism among the earliest Greeks, he concludes: "One needs to have a singularly biassed mind, to attach the least value to them. There is not, in the whole of ancient literature, a single passage which, sanely interpreted, confirms such an assertion."[1]

If we adhere to Homer and Hesiod, we find that all ownership is private, so far as arable land is concerned. The estates of Alcinoos, of Odysseus, of Perses, are private properties; nowhere is there any mention of cultivated land owned collectively. Here are two passages, from which Esmein tries to argue the existence of collective ownership. On the shield of Achilles, the divine smith has represented a vast piece of fallow which many workers are ploughing. When the ploughs come to the ends of the furrows, a person, whom the poet calls *ἀνὴρ ἐπιών*, gives every ploughman a cup of wine, and then work starts again.[2] It is a falsification of the meaning of the lines to regard this ground as "land of the community divided into equal portions allotted to individuals to work."[3] Elsewhere the poet likens the Greeks and Trojans, facing each other from either side of a rampart, to two men who, measure in hand, on a piece of common land, dispute every inch in order that their two portions may be equal.[4] Here we have a perfect picture of collective property, says Esmein.[5] One must really be the slave of a preconceived idea, to interpret the scene in this way. On the contrary, it seems to me that the attitude of the two neighbours bears witness to the existence of private property, and to the stubbornness with which each fought for his own portion.

Moreover, the first care of a man founding a city is to divide up at least part of the land among his comrades. Nausithoos, we read in the *Odyssey*, settled the Phæacians in the island of Scheria; there he founded a city, built houses, consecrated temples to the immortal gods, and proceeded to distribute the land.[6] "At the birth of most cities, the initial

[1] **XLV**, pp. 21 *ff.*
[2] *Il.*, xviii, 541 *ff.*
[3] Esmein, *art. cit.*, p. 834.
[4] *Il.*, xii, 421 *ff.*
[5] *Art. cit.*, p. 833.
[6] vi, 7 *ff.*

act which we find is a distribution of the land among the citizens. The public authority did not thereby create the concept of property, which was much earlier than the State, but it created proprietors."[1]

Whatever may have been claimed to the contrary, the stories of Meleager in Ætolia and of Bellerophon in Lycia do not, any more than the speech of Achilles to Æneias, tell against the existence of private property. Here it is a matter of lands promised or given to individuals by cities in special circumstances. There is no justification for supposing that these lands had previously been assigned to individuals, and that therefore the State was still the true owner of them, since it conceded them to others. It is much simpler to conclude that cities possessed common land, part of which they could make over as gifts. There is no contradiction between the existence of such land and that of private property. These are two forms of ownership which can quite well be found side by side.

But was this private ownership, which I consider to have been the usual manner in which arable land was held, individual ownership or family ownership? Neither the *Iliad* nor the *Odyssey* gives definite information on the point. Alcinoos and Odysseus own big estates. When Odysseus is away, it is Telemachos who sees that his father's land is worked properly and endeavours to defend it against the greed of the suitors. But there is no indication to enlighten us as to the true meaning of that right of ownership. Does the property belong to Odysseus and to Telemachos as individuals, or are father and son merely the successive representatives of the family, the *genos*? The Homeric poems are silent on this point. Hesiod, on the other hand, is quite clear. Speaking to his brother Perses, he reminds him that the landed property left by their father, the *kleros*, had been divided between them, that the division gave rise to a lawsuit, and that Perses, to obtain the better portion and to win the case, bribed the judges.[2] Here, then, we certainly have individual, not family, ownership. Each of the two sons, on the father's death, becomes or should become the possessor of part of the property (*kleros*), which was previously one single estate.

The very definite indications given in this passage of

[1] XLV, p. 34.

[2] *Works and Days*, 37 *ff*.

Hesiod allow one to ascribe a historical meaning to two passages in the *Iliad* and *Odyssey*. In the *Iliad*, Poseidon says that the universe was divided between the three sons of Cronos and Rhea, Zeus receiving the dominion of the sky, Poseidon that of the waters, and Hades that of the underworld.[1] In the *Odyssey*, Odysseus tells Eumæos that he is the bastard son of a wealthy Cretan, whose goods were divided up after his death, and that he himself inherited only a house and a very small portion.[2] In both cases the portions were assigned by lot.

It is true that there is much evidence to show that inheritances were not thus divided always and everywhere.[3] We must conclude that inheritances were or might be transmitted undivided in practice, but that there was no law that they must be. There is nothing to prevent us from assuming that in Homeric and Hesiodic society there was not only private, but individual, property.

So far, we have chiefly considered the ownership of arable land. Was it the same with pasture? Here again one must beware of general and absolute statements. M. Guiraud has observed that in describing most of the estates mentioned in the *Iliad* and *Odyssey* the poet speaks of plough-land, orchards, and vineyards, but not of meadows or grazings. He notes, however, that the pasture-land on which the countless herds of Odysseus graze is the hero's own property, "for his slaves have built fine byres of unhewn stone on it, surrounded with a thorn hedge."[4] The author of the *Works and Days* advises his brother Perses to get the fodder and bedding into the barn as soon as the harvest is in.[5] The bedding may consist of the straw of the ingathered corn, but the fodder must of necessity come from meadows entirely owned by Perses or rented by him as an individual. A private individual could not supply himself with fodder from common land. It may, on the other hand, be allowed that the mountain-sides and the undergrowth of the forest, where flocks and herds were taken to graze, had not been divided among the members of the community, and that all had the right to use them.

So the system of landed property, in the society portrayed in the Homeric and Hesiodic poems, is far from being uni-

[1] xv, 187 *ff*.
[2] xiv, 208 *ff*.
[3] **XLV**, pp. 55 *ff*.
[4] *Ibid*., p. 68.
[5] 606.

form. We find private ownership and individual ownership, without being able to say that there was no family and collective ownership at all. The reality was more complex than the theories constructed by modern erudition lead one to suppose. The facts to be found in the texts bring out that complexity. There is no use in trying to twist these facts to make them suit preconceived ideas.

VI

Industry

By the side of agriculture, stock-breeding, hunting, and fishing, industry had a big place in the economic life of Homeric society—not, indeed, industry on a large scale, as organized in the modern world, but industry in the general sense of the word, that is, the transformation of various raw materials into things intended to satisfy men's needs or luxurious tastes.

The raw materials known and used by the Greeks of this period were building-materials (stone and marble), metals (copper, tin, iron, silver, and gold), textiles (flax and wool), wood, hides, and clay. We are given no details about stone and marble quarries or metal-mines, nor do we know whether the Greeks grew their own flax. Wool and hides were supplied by the beasts which they reared and the game which they loved to hunt, and they found plenty of wood in the forests which covered their country.

The palaces described in the *Iliad* and *Odyssey*, those of Priam, Odysseus, Nestor, Menelaos, and Alcinoos, bear witness that huge buildings existed, lavishly adorned with metal ornaments. The building-industry was prosperous. Dikes and bridges were also built, to say nothing of private dwellings, provided with barns and byres, like those which Eumæos caused to be erected.

Metal-working already reigned supreme. The common and precious metals were worked by smiths, armourers, and goldsmiths with a lavishness, a technical skill, and a decorative sense which astonish one. It is enough to mention the shield of Achilles in the *Iliad*, the belt of Heracles in the *Odyssey*, the shield of the same hero in Hesiod, the arms of Agamemnon in the *Iliad*, and the many vases, bowls, tripods, and craters of bronze, silver, or gold, which are mentioned, to show the place taken by the metal crafts at the time. The smiths worked, as today, with bellows, anvils, hammers, and tongs. The goldsmiths did chasing and hammer-work, set stones, did gilding and silvering, and combined amber and ivory with gold.

The textile industry consisted in the spinning and weaving of flax and wool and the manufacture of clothes, embroideries, and carpets.

Hides, tanned and dressed, served as cloaks for the men. Fixed together in several layers, they were used to make shields. Shoes were made of leather.

The wood industry was highly developed. The logs were brought from the forest by teams of mules, and cut up into thick beams and planks, which were used for heavy timber-work and joinery. The rule and the compass were used in building boats and making furniture. Farm-gear, such as pestles, mortars, beetles, and ploughs, was made of wood.

The potters used the wheel for making common jars and vases.

The organization of all this industrial labour took two forms. There were household crafts and specialized crafts. Weaving, cooperage, leather-work, and wood-work were done at home. In every family, the women, free and slave, span, wove, made clothes, and did embroidery and tapestry. Eumæos cut his leather himself and made his own shoes. Odysseus was a marvellously skilful carpenter and joiner; he made his bed out of the trunk of an olive-tree and with his own hands built the boat which was to take him away from Calypso.

In metal-working, and doubtless also in building and pottery, labour was specialized. In addition to the divine smith Hephæstos, the *Iliad* mentions a celebrated armourer named Tychios, and in the *Odyssey* the craftsman Laërces is bidden to gild the horns of the heifer sacrificed by Nestor in honour of Athene. The poet of the *Works and Days* mentions the builder (τέκτων) and the potter (κεραμεύς) in a list of various trades.

But, active and skilled as this industry already was, it only seems to have worked for local and immediate needs. Nowhere do we find the slightest trace of any production organized for commercial purposes. The articles manufactured do not appear to be used for exchanges between the Greeks themselves or with foreigners.

VII

Trade. Exchanges. Voyages

This does not mean that trade was unknown in the society of the day, but for the Greeks it consisted essentially in importation. Its field of action was fairly small; its development was restricted by the methods of exchange.

Within Greece itself, between one district and another, between one city and another, trade relations grew up. When Pallas Athene appears for the first time before the eyes of Telemachos, she presents herself in the form of Mentes, chief of the Taphians. " I am going," she says, " with a ship and a large crew to another land, to Temese; I am taking iron there, to exchange for bronze."[1] Among the questions which Nestor asks Telemachos when he arrives at Pylos, we find, " What interest, what business bring you ?"[2] Hesiod advises Perses, if he wants to do a little coast-trade, not to be too ambitious, and to load his boat with only part of his harvest.[3] Trade of this kind could only be between districts very close to one another. Indeed, in respect of the agricultural and industrial products of Greek lands we may take it that at this time every separate district in Greece was almost self-sufficing.

But Greece did resort to foreign goods, or rather it was not indifferent to the foodstuffs, the raw or precious materials, and the manufactured articles which foreign mariners brought to its doors. The Greeks liked the wines of Thrace. We read in the *Iliad* and *Odyssey* of purple veils of Sidon, and of goldsmith's work from Phœnicia—vases, tripods, silver baskets. Copper, tin, amber, and ivory must have come from abroad. The Greeks also bought foreign slaves. Eumæos tells Odysseus that there had been a Phœnician slave-girl in his father's palace, and a Sicilian bondwoman serves in the house of Laërtes.

In payment for these purchases the Greeks seem to have

[1] *Od.*, i, 182 *ff*. [2] *Od.*, iii, 72.
[3] *Works and Days*, 618 *ff*.

chiefly given cattle. Among the prizes which Achilles proposes for the funeral games in honour of Patroclos are a tripod valued at twelve oxen and a slave-girl valued at four oxen. Laërtes paid twenty oxen for Eurycleia. When the author of the *Odyssey* speaks of the purchases made by the Phœnicians in return for their sales, one must doubtless understand the cattle given to them in exchange for the metals or precious articles which they have brought. From this point of view, one can just speak of Greek exports, but they were much rather a payment in kind, mere barter.

The countries with which Homeric and Hesiodic Greece thus entertained commercial relations, direct or indirect, were chiefly grouped round the Eastern Mediterranean; so far as one can judge from the poems, they were Thrace, Asia Minor, Cyprus, Phœnicia, and Egypt. From where did the amber come ? Two hypotheses are possible. Amber existed nowhere except on the coasts of the Baltic, and it was apparently brought by land either to the north coast of the Euxine or to the top of the Adriatic, where it was picked up by Phœnician vessels which conveyed it to Greece. Ivory might come from Central Africa, through Egypt, or else from India, over Iran and Western Asia, and the Phœnicians received it in their own cities or went to the Nile Delta for it, as the case might be. Did tin, at this early date, come from the Cassiterides ? The Western Mediterranean and the Atlantic were still almost unknown waters to the Greeks, but the Phœnicians may have found their way there.

From all the pieces of information contained in the Homeric and Hesiodic poems one gathers that the Greeks practised coastal navigation, but did not go far from their own shores. Long sea-voyages were quite exceptional; there was Menelaos, who, on the way back from Troy, went to Cyprus, Phœnicia, and Egypt, and there was Odysseus, tossed by storms over the Western Mediterranean and drawn on to the dark land of the Cimmerians by his desire to speak with the dead. In any case, these two voyages are really quite as legendary as that of the Argonauts, which took in Colchis, the Pillars of Heracles, and Libya. They cannot supply us with serious evidence on the extent of Greek commerce in those distant days. The purveyors of the Hellenes at that time were the Phœnicians, at once merchants and pirates, sea-traffickers who had taken the place of the Ægeans in the Eastern Mediterranean as they would themselves one day be supplanted by the Greeks. Hardly has Odysseus set foot on Ithaca when he tells Pallas Athene, not recognizing her, that Phœnicians have brought him from a Cretan port to his own country.

Whatever the character and extent of this trade may really have been, what gives it its special aspect is the absence of any money in the transactions of which it consisted. We have already seen passages in which the value of a slave or a tripod is expressed by a certain number of oxen, and we have seen the lines in the *Odyssey* in which the supposed Mentes declares that he is going to Temese to barter iron for bronze. Purchase and sale are nothing but exchanges in kind, in which the two principal elements of the operation are, on the part of the Greek buyers, livestock and metals, not minted, but raw, and not only precious, but base metals. Livestock and metals are their chief form of wealth. When Eumæos would give a catalogue of the wealth of Odysseus, he enumerates twelve great herds of oxen, twelve great flocks of sheep, and as many pigs and goats, which his master owns on the mainland, and eleven great herds of goats and hundreds of hogs and swine on Ithaca.[1] What are the gifts which Iphidamas promises to his young wife ? A hundred oxen, a thousand goats, and a thousand lambs.[2] The poems contain very many instances of wealth reckoned in flocks and herds. Metals, too, have an important place in primitive fortunes. In the story which Odysseus tells Eumæos of his imaginary adventures, he passes himself off as a Cretan who has travelled far, and has been cast up by a storm on the coast of the Thesprotians, where he has had news of Odysseus.

> "The King of the Thesprotians," he says, "told me that he had treated him kindly, and as a guest. He showed me all the possessions which Odysseus had amassed, bronze and gold and wrought iron. Any other would have had enough for ten generations, so much wealth was there, laid up in the halls of the King."[3]

At Ithaca itself, the real treasures of Odysseus consisted of stuffs, clothing, bronze, gold, and wrought iron.[4]

The prizes which Achilles proposes for the funeral games of Patroclos are vases, tripods, beautiful slave-girls, mares or

[1] *Od.*, xiv, 100 *ff.*
[2] *Il.*, xi, 244 *ff.*
[3] *Od.*, xiv, 321 *ff.*
[4] *Od.*, xxi, 10.

mules, bronze, gold, and iron. The details which the poet gives about this last metal are very interesting. For the disk-throwing, the prize offered to the winner is the disk itself, a mass of unwrought iron, and Achilles adds:

> "Even if he have many rich lands, this piece of iron will be enough for his needs for five whole years. His shepherds and ploughmen will not need to go to town for lack of iron; it will be there."

No less curious is the prize for the pigeon-shooting.

> "And he gave dark iron as the prize for the bowmen, and set down ten two-bladed axes and ten half-axes."[1]

Prehistorians have supposed that the deposits of bronze axes, so frequent in the Bronze Age, were, at least in some cases, treasures hidden in the ground; the passage in the *Iliad* seems to confirm the supposition, bringing the custom down into the Iron Age.

But, while the metals, precious, like gold and silver, or useful, like bronze and iron, constituted at least a part of wealth, coining was completely unknown. When "talents of gold" are mentioned, weighed ingots are meant. One of the prizes for the running in the funeral games of Patroclos is a half-talent of gold.[2] To bring Achilles back to fight, Agamemnon offers him, in addition to tripods, precious vases, and horses, ten talents of gold.[3] Here the talent is a unit of weight, not a monetary denomination. Already, however, the part played by the metals in trade and in private wealth marks, as it were, the transition from the age of barter pure and simple to that of coinage.

In the society depicted in the *Iliad*, the *Odyssey*, and the works of Hesiod, economic activity is mainly agricultural and pastoral. Wealth comes almost solely from the cultivation of arable land, cornfields, vineyards, orchards, and gardens, and the exploitation of pastures. It is by sacks bulging with flour, jars full of wine, and head of cattle that the wealth of chiefs and great men is reckoned. That opulence is displayed in vast dwellings, sumptuously furnished and decorated.

[1] *Il.*, xxiii, 830 *ff.*, 850 *ff.*
[2] *Il.*, xxiii, 751.
[3] *Il.*, ix, 122.

Industry serves only to supply necessities and luxuries. It is not an independent form of human labour; it does not aim at the rational exploitation of the many and various raw materials which nature has to offer man. The industries connected with food, clothing, wood, hides are still purely domestic in character. In every household, flour, wine, and oil are made, beasts are slaughtered for the table, materials, fine or coarse, are spun and woven for clothes, carpets, hangings, and embroideries, and leather and wood are adapted to various purposes. Only the metal and pottery industries appear to be specialized, doubtless on account of the material equipment and premises which they require. Also, one should observe that most of the metals—gold, silver, copper, tin—are very probably imported from abroad, and so are many artistic objects—precious vases, jewels, etc.

Trade is very little developed. The Achæans do no real exportation. A few exchanges take place between district and district, between city and city, within Greece itself. The Phœnicians, the pedlars of the sea, land in harbours and on beaches, and there sell the products of their own industries, or foodstuffs, raw materials, and manufactured goods which they have fetched from all the shores of the Mediterranean and the distant lands of the East. All this trade is done entirely by barter. Money is unknown. Tripods or slave-girls are given in exchange for cattle, iron for bronze.

Neither industry nor trade is a source of profit to the Greeks. The society of the period lives chiefly by farming. It is obstinately sedentary. Only serious, abnormal causes determine men to travel. Every man's life is bounded by a very close horizon. Navigation is still unadventurous and infrequent. But the day is not far off when, driven by urgent necessity, Greece will come out of herself. Then her economic life will be transformed and will rise to its full development.

CHAPTER II

THE EXPANSION OF GREECE OVER THE MEDITERRANEAN[1]

FROM about the eleventh century to the end of the seventh, the Greeks spread over the Mediterranean in every direction. They founded numerous colonies on the Eastern and Central Mediterranean, they reached the Euxine by the Hellespont and Bosphorus, and they made their way beyond the Strait of Messina to the Western Mediterranean. In addition to the old Greece, which had been confined to the southern part of the Balkan Peninsula, there were now, among other territories, an Asiatic Greece, covering the Ægean seaboard of Asia Minor, and a Western Greece, the cities of which were collected in the south of Italy and in the greater part of Sicily. The Ægean Sea and the Ionian Sea were embraced in the Hellenic world.

This expansion of the Greeks, this creation of colonial Greeces outside the original Greece, is a very big historical fact, which had a great effect upon the economic life of Greece and of the whole Mediterranean. The Greeks not only caused their political and social institutions, their ideas, their arts, their civilization, to radiate wherever they settled. By their activity and labour, they transformed the economic conditions of many countries. They came into direct contact with some of the states of the East, with regions which produced foodstuffs and raw materials, and these contacts affected their own economic life.

The colonial expansion of the Greeks was a long, varied, and complex process. Its causes were many. We find streams of migration, sometimes parallel, like those crossing the Ægean from west to east, and sometimes divergent, as when the Greeks left the Ægean by all its outlets, north-east,

[1] **XXVI**, vol. i, chaps. ix-x; **XXVIII**, vol. i, chap. ii; **XXIX**, vol. i, chap. ii, §§ 8 *ff.*; **XXX**, vol. i, pp. 142 *ff.*, 507 *ff.*; **XXXIII**, pp. 51 *ff.*; **CV**, no. 5.

south-east, south, and west. There was no more uniformity in the character of the settlements or in the economic consequences of the movement than there was in the causes and general directions of the movement.

This is not the place to set forth in detail the history of that colonization, and an admirable summary of it will be found in M. Jardé's work on *The Formation of the Greek People.*[1] But we should dwell on the facts which are of especially economic interest; we must try to discover the economic causes and consequences of these migrations and to estimate their importance and all that they entailed.

I

The Economic Causes of Greek Colonization

Several ancient authors, Thucydides[2] and Plato[3] in particular, state that the cause of the departure of the Greek settlers was lack of land—that is, an economic fact. But why was there a lack of land ? It does not seem possible to ascribe it to over-population.[4] M. Guiraud holds that the principal, if not sole, cause of it was the organization of property, which in his opinion belonged to the family, not to the individual, and could not be divided among brothers. I have tried above to show that, according to the Homeric and Hesiodic poems, this theory at least goes too far, and that in the Greek society depicted in the *Iliad*, *Odyssey*, and *Works and Days* there were individual properties, and that inheritances of landed property could be divided. Moreover, emigration can equally well be explained by the dividing up of a father's estate among many legatees, for if each portion thus obtained was too small to maintain a new household many individuals would be compelled to go abroad in search of bigger lands. M. Guiraud also speaks of the "breakaway," as he calls him, the man who, willingly or otherwise, has broken with his family and left it, from a desire for independence, wealth, or adventure, or else because he has committed a crime. The examples which he quotes, those

[1] **XLVII**, English pp. 171-228.
[2] i, 2. [3] *Laws*, iv, 708B; *cf.* 707E.
[4] **XLV**, pp. 79 *ff.*; *cf.* **XLVII**, English p. 180.

of the Pylian Melampus and Theoclymenos, are quite exceptional, and do not explain such a general fact as Greek colonization. Here again the reasoning is based on comparisons, in my opinion quite inapplicable, from early Ireland, primitive Germany, and Central India.[1]

As a fact, the first exodus of the Greeks was provoked by spoliations, apparently due to various causes. The movements of peoples known under the name of the Dorian invasion, which especially affected southern Thessaly, Bœotia, and the Peloponnese, brought about a collision between the old possessors of the soil and the conquering new-comers in those parts. Later, we shall see what effect the struggles which then disturbed Greece had on the organization of landed property. Here we must merely note that among the conquered there were many who, rather than accept a degrading or at least diminished status, left their country to look for lands which they could make completely their own. So bodies of men set forth from southern Thessaly and Bœotia, and founded the Greek cities of Æolis. Among the invaders themselves there were some who, perhaps less favoured than others and dissatisfied with their share of the spoil, perhaps more adventurous, too, continued their advance over the sea. Not only Achæans, but Dorians, crossed the southern Ægean to Cos, Rhodes, and the long, narrow capes and headlands of Caria. In the midst of all this shifting of populations, Attica, being almost untouched, saw her meagre soil invaded by more refugees than she could feed, and from her shores the Ionian colonization reached the Cyclades and Asia Minor.

The civil wars and revolutions through which many cities passed before they gave themselves fairly stable political institutions often ended in the confiscation of the land owned by the leaders and chief men of the defeated parties, who were, moreover, sent into exile, willingly or by force. These too must have gone and settled in a distant country where, on a new, free soil, they once more found the property which they had lost at home.

In addition to these general causes there may have been strictly local or individual reasons in certain particular cases,

[1] XLV, pp. 81 *ff*.

and Seneca seems to have put together in a single sentence all the knowledge which the ancients possessed on the subject:

> "*Alios excidia urbium suarum hostilibus armis elapsos in alienas spoliatos suis expulerunt. Alios domestica seditio submovit. Alios nimia superfluentis populi frequentia ad exonerandas vires emisit. Alios pestilentia aut frequentes terrarum hiatus aut aliqua intoleranda infelicis soli vitia ejecerunt. Quosdam fertilis orae et in majus laudatæ forma corrupit.*"[1]

If lack of land, due to various circumstances and accidents, was what caused the earliest Greek colonists to emigrate, other factors came later to play a larger and larger part in the progress of Hellenic expansion. First necessity and then experience and a love of the life led the Greeks to take to seafaring. Many colonies in Asia, clinging to the often mountainous coast and prevented from acquiring a large territory by the rugged nature of the interior or the unconquerable hostility of the tribes which dwelt there, found themselves compelled to turn their economic activities seawards. When the Hellenes took the place of the Phœnicians in the traffic among the islands and coasts of the Ægean, they were obliged to go themselves to the producing countries, or as near to them as possible, to seek the foodstuffs, raw materials, and manufactured goods which, for several centuries, the ships of Tyre and Sidon had brought to Greece. Many Greek colonies, on the various coasts where they were planted, entered into direct contact and trade with peoples, some of which were civilized and others still half-barbarous. As this trade developed, with the sales, purchases, and transport which it involved, new markets were naturally sought, and further posts were set up from which there was easy access to the producing countries; that is, still more colonies were founded.

All these causes, which, especially in the eighth and seventh centuries, were operating simultaneously, helped to give the Greek movement of expansion great economic importance. Not only Greece and the Greek race won profit and greatness from it; the economic life of the Mediterranean underwent a profound and lasting change.

[1] *Consol. ad Helviam*, vii, 4, quoted in **XLV**, p. 78 n. 1.

II

The Economic Character of the Greek Colonies

The variety and development of the economic causes of colonization explain the diversity in nature and activity of the innumerable colonies founded by the Greeks.

The bodies which left Greece to look for land were chiefly anxious to obtain arable and pasture land of some extent. Preferably they chose places with a climate like that of the mother-country, and therefore likely to be favourable to the crops which they were used to growing at home. Such, at the beginning at least, was the character of the Ionian colonies of Asia Minor, at the mouths of the valleys which descended westwards from the Anatolian plateau, and in particular those of the Hermos, Cayster, and Mæander. The territories of these cities were rich in cornfields, pasture-land, olive-groves, and fruit-orchards. Cyrenaica very soon won a name for great agricultural wealth; corn-growing and stock-breeding throve there. In Great Greece and Sicily, the immigrants from the Peloponnese, Corinth, Megara, Chalcis in Eubœa, and Locris found fertile plains, on which wheat, vine, and olive yielded almost miraculous crops and cattle, large and small, found abundant grazing. In these countries, especially in Cyrenaica and the West, the old population was thinly distributed or still semi-barbarous. The Greeks brought their activity and their experience in agriculture and stock-breeding, and rapidly developed an intensive exploitation of all the pastoral and agricultural resources, favoured at an early date by increasing exportation.

Elsewhere, in Chalcidice, Thrace, and Cyprus, the Greek colonists went in search of resources of another kind—timber and the precious and useful metals, such as the gold of Thrace and the copper of Cyprus.

Besides these settlements, whose economic life, whether agricultural, pastoral, or, as in some cases, industrial, was based on the direct exploitation of the soil, the fruits which it bore and the raw materials which it contained, there were others of a more specially maritime and commercial nature. At an early date the Greeks perceived the importance of the Hellespont and Bosphorus. To secure a free passage for themselves and also, doubtless, to assist navigation, they established "ports of call and watch-posts"[1] on the Asiatic and European shores—Sestos, Perinthos, Byzantion, Abydos, Lampsacos, and Chalcedon. On the Euxine, they established themselves on the ends of the natural roads which ran from the Danube valley, from the vast plains of southern Sarmatia, from Caucasia, and from Armenia. These Greek colonies dotted along the coast were trading-stations, centres of sale and purchase, importation and exportation, which for a long time supplied Greece Proper with the foodstuffs and raw materials which she lacked and the native peoples with many products of Greek industry, especially pottery and jewellery.

In the Delta of the Nile, Naucratis was a veritable entrepôt. Beyond Great Greece, on the distant coasts of Gaul and Spain, Marseilles (Massalia) and her dependencies, from Nice to Rosas (Nicæa, Rhode), were mainly trading-stations, playing the same part in the Far West as was played by the colonies of the Euxine in the north-east and Naucratis in the south of the Hellenized Ægean.

There is no doubt that there were two distinct types of Greek colony, whose respective natures were determined by the causes and conditions of the migrations. But it would be an over-simplification and a positive distortion of the history of that expansion to maintain this point of view absolutely. Here, as everywhere and always, time did its work, and there was development. One colony, founded by a body of emigrants who were looking for land, and at first living by agriculture and stock-breeding, would presently enter into trade with the state or tribes whose territory marched with its own. Another, standing at the end of one or more important routes, long in use, would become the Mediterranean outlet of countries, often far away, whose products found favour with the Greeks. Another, being within reach of raw materials, textile, metallurgical, or ceramic, would gradually develop an industrial activity, with which it would feed its trade. Such was the story, to quote only the most decisive instances, of Miletos, Cyrene, Syracuse, and Taras.

The settlements which were founded chiefly to do trade with the barbarians must not be regarded as mere "factories," like those set up by the Phœnicians in Greece in the preceding

[1] **XLVII,** English p. 204.

period. The colonies of the Euxine in the north-east and Marseilles in the west were deeply rooted in the soil, of which they owned and cultivated a considerable extent. It was from there, no doubt, that their influence radiated. But their economic activity did not consist solely in trading and shipping. They also exploited, by their own labour, the arable land, pastures, forests, and other resources of the surrounding region.

III

The Economic Consequences of Greek Colonization

The immediate consequence of this expansion of the Hellenes, what at once strikes the attention of the historian, was the great increase of the geographical area henceforward covered by the economic activity of the Greek race. From being confined to Greece Proper and the immediately neighbouring islands, that activity now extended to many vast regions, across the Ægean, round the Euxine, in Libya, in Sicily, in southern Italy, and as far as Cyprus in the east and Gaul and Iberia in the west. When the Greeks started to work the natural resources of a country the value of the land increased considerably. The extent and the return of arable, prairie, and pasture were greatly increased; the variety and abundance of raw materials of all kinds brought a new element of prosperity; the circle of trade relations widened to all sides of the horizon.

Then the part played by the sea in the national life of the Greeks took definite shape and rose to its full importance. By the sea, and only by the sea, the colonies communicated with their mother-cities. Their independence, political and economic, and their very existence demanded a strong navy. Corinth, Chalcis in Eubœa, Miletos, Phocæa, Rhodes, Syracuse, Taras, and Marseilles had great fleets, mercantile and armed, long before Athens became mistress of the Greek seas. So, at the same time as the Hellenes were occupying new lands, almost all very rich, outside their native country, and were creating centres of influence and business on many a distant shore, they were taking possession of the sea which at once divided and connected all parts of the Greek world.

The result of this mastery of the waves and the sea-routes

was that the economy of Greece shook off the once undisputed dominion of the Phœnicians. The old position of that people as middlemen between the great empires of the East and a Greece which was, as it were, tied to its own soil, came to an end. Greek ships went to the end of the Euxine for the metals of the Caucasus and Armenia; Greek ships brought to the ports of Hellas the foodstuffs, raw materials, and manufactured goods bought in Cyprus, Syria, and Egypt; Greek ships carried all the products exported by Ionia, Cyrenaica, Great Greece, Sicily, and the colonies of the Western Mediterranean.

The mainly rural economic life described in the Homeric and Hesiodic poems was henceforth supplemented by a growing industrial activity and an ever bolder commercial expansion. It was not only for their own domestic and immediate needs that the Greeks now worked flax and wool, metal, wood, clay, skins, and leather. All these manufactures, destined to feed a national and international trade, developed and became important factors in economic life. In the past, the Phœnicians had landed their goods at the Greek anchorages and had exchanged them against native products—usually, it seems, cattle. Henceforward the Greek mariners would themselves go to Egypt, to Syria, to Asia Minor, and among the peoples of Europe, the civilized Etruscans and the barbaric Scythians, Gauls, Ligurians, and Iberians, taking with them manufactured goods and works of art, tissues, weapons, jewellery, and painted vases, which had a great reputation and were eagerly bought by all the barbarians.

It was thanks to their colonial expansion that the Greeks became acquainted at an early date with that admirable instrument of business, money. Movable wealth then took its place by the side of landed wealth in the economy of the Greek world. "In such a matter," says M. Glotz, "all is action and reaction. The development of trade has made a practical and certain instrument of exchange necessary, and the invention of money contributes mightily to the expansion of trade."[1]

If, then, we confine ourselves to observing, within the limits of the widened Hellenic world, the effects of the colonial movement, we note an unprecedented development of

[1] **XLIV**, English p. 69.

all forms of economic activity. Agriculture, industry, and trade are practised with an intensity hitherto unknown. From the soil, from workshops, from ports, and from trading-stations, flows a stream of wealth which not only swells incessantly, but spreads, is diversified, is transformed, and contributes, in its turn, to increasing the number and the productivity of the sources from which it comes.

The whole Mediterranean takes on a new economic aspect. In many regions round the central sea, the land, being better worked, is covered with rich crops, as in Cyrenaica, Sicily, and southern Italy. On shores previously unknown or barely known, commercial activity springs up; so it is on the northern coast of the Euxine and on the seaboard of Gaul and Spain. The Phœnicians, once the masters of sea-borne trade between East and West, are driven by the Greeks out of the Ægean, the Euxine, and the Ionian Sea. They keep their naval supremacy only on the African coast, west of the Greater Syrtis, and in the region of the Pillars of Heracles. The Tyrrhenian Sea witnesses the struggle of the Greeks against the Carthaginians and Etruscans. The keen rivalry of the sea-faring and colonial powers gives a lively impulse to navigation.

By the beginning of the sixth century, the economic life of the Hellenic world and of the Mediterranean world has assumed the general character which it will have until the expedition of Alexander the Great. The ups and downs of Greek history—civil conflicts, wars with the Persians in the east and with the Carthaginians in Sicily—will not, any more than the events which brought Rome and the Latins and Etruscans into collision in Central Italy, make any appreciable difference in the main features of that life.

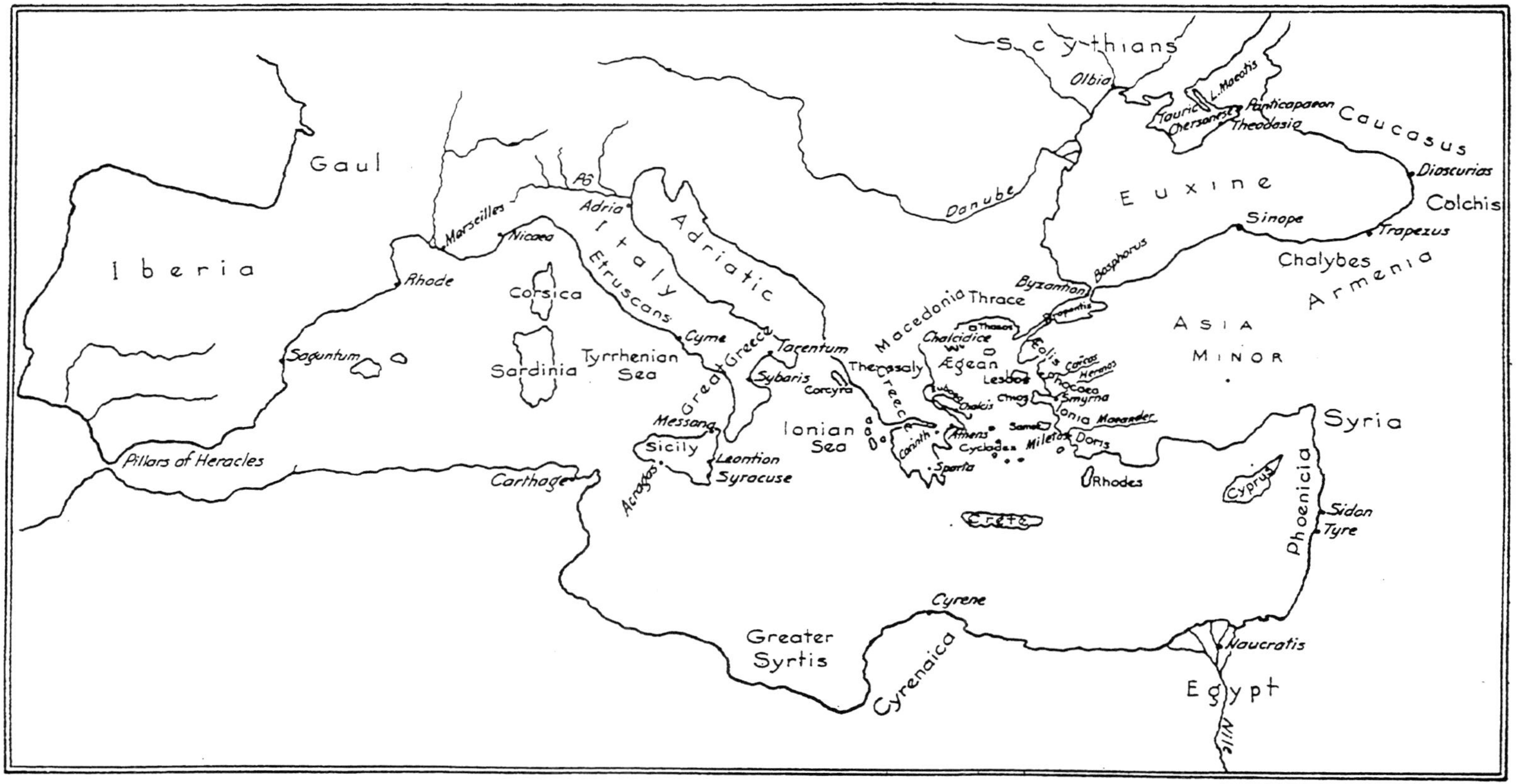

MAP I. THE EXPANSION OF THE GREEKS OVER THE MEDITERRANEAN.

[face p. 32.

CHAPTER III

AGRICULTURE AND AGRICULTURAL LIFE. RURAL PROPERTY IN THE GREEK WORLD FROM THE SIXTH TO THE FOURTH CENTURY BEFORE CHRIST

ONE economic consequence of the colonial expansion of the Greeks was that the area of land, tilled or otherwise exploited by them, was very greatly increased. The new territories were among the most fertile in the Mediterranean region—the greater part of Sicily, the best parts of southern Italy, the plateau of Cyrenaica, the plains and low-lying valleys of western Asia Minor, the seaboard of Thrace, and the shores from the Hellespont to the Bosphorus.

Agriculture also benefited by another characteristic process which was going on. The forests, which had once been very big, dwindled in extent. The Greeks had a great need of wood, for building, carpentry, and shipbuilding. Deforestation left hill-sides free for cultivation. Like the forests, the pastures also gave place in many parts to plough-land. The turning of prairies and underwood where herds used to graze into arable had the same effect as deforestation.

Colonial expansion, deforestation, and reclamation of hitherto untilled soil—these three facts helped to determine the chief features of agriculture and agricultural life in the Greek world from the sixth to the fourth century B.C.

The consequent extension of cultivable land was not used solely, nor even chiefly, for the benefit of corn-growing. At an early date the Greeks preferred to develop the growing of trees, and first and foremost of the vine and olive. Lawgivers and heads of states, Solon,[1] Peisistratos,[2] Gelon of Syracuse,[3] recommended and encouraged the planting of vineyards and orchards. This action on the part of individuals and some governments is commonly attributed to a desire for gain, on

[1] Plut., *Solon*, 23. [2] Dion Chrys., *Or.*, xxv, 3.
[3] Plut., *Apophth.*, Gelon, ii.

the ground that the profit made from exporting wine and oil was larger than the cost of importing corn. There may have been another reason, in the climatic conditions which prevailed in the Greek world as a whole. All round the Mediterranean, the rain falls seldom, but when it does it comes in torrents, and is followed by a heavy flow of water on the surface of the ground. On treeless slopes this surface flow, not being checked or slowed down by any obstacle, sometimes carries everything away with it. The presence of trees lessens the violence of atmospheric precipitations, and the humus collected round their roots, increased every year by the fall of the leaves, absorbs part of the rain-water and reduces the erosion of the soil. Tree-growing, therefore, is a remedy for the disadvantages of the Mediterranean climate, and the development of it may have been regarded by experienced men as a compensation for the loss of the forests.

However this may be, and whatever the real reasons may have been, vineyards and orchards were of very great importance in the agricultural economy of ancient Greece. Lease-contracts sometimes stipulated that the lessee should plant a given part of the estate in question with fruit-trees.[1]

I

Agriculture, Stock-Breeding, Forests, Hunting, and Fishing

The Greeks looked to the earth for three kinds of vegetable food—corn, fruit, and vegetables.

The kinds of corn far the most cultivated were barley and wheat. The Greeks do not appear to have known rye or oats. They grew a little millet.[2]

The trees most extensively grown were the vine, olive, and fig. The manufacture of wine and oil and the drying of figs took up a large part of the Greek peasant's time if the harvest was good. The other trees of our own regions, bearing fruit with stones or pips, were also known to the Greeks, but none of them had the importance of the fig, olive, and vine.

There were many and various vegetables; a list of them

[1] **III**, no. xiii *bis* (pp. 239 *ff.*), §§ 1, 4.

[2] **XLVIII**, *passim*.

will be found in Daremberg and Saglio's *Dictionnaire des Antiquités grecques et romaines*, in the article "Cibaria."[1] Textile plants, except perhaps flax, seem to have had no place in Greek agriculture.

Special mention should be made of a plant which still remains a mystery, in spite of all efforts to identify it—the famous silphium of Cyrenaica. It is certain that the Greeks cultivated it, but at the end of ancient times it had almost disappeared from Cyrenaica, and it had to be sought far south, not without danger. It was used for food, medicine, and perfumery.[2]

Of all these products, only the corn was not grown in a sufficient quantity for the needs of the Greeks. The importation of wheat was in many Greek cities, and especially in Athens, a necessity of the first order. Wine and oil, on the other hand, were exported largely.

The various regions of the Hellenic world were not all equally favourable to these different crops. Corn, vines, and fruit-trees all grew together in some parts—in Greece Proper, in the alluvial plains of Thessaly and Bœotia and in the valleys of the Eurotas in Laconia, the Pamisos in Messenia, and the Alpheios in Elis, which opened widely towards the sea; in Asia Minor, in the lower valleys of the Hermos and Mæander; in Cyrenaica, in the parts of the plateau nearest the sea; and in the west, in Sicily and the plains surrounding the Gulf of Taras and a few small districts on the Tyrrhenian shore south-west of Cyme (Cumæ). Attica, though poor in wheat, was renowned for its wine and oil. Phocis, Ætolia, Acarnania, and Arcadia, being chiefly mountainous, allowed of little cultivation, but abounded in forests and pastures. The islands of the Ægean and the coasts of Caria, Æolis, and the straits leading to the Euxine had no ground suitable for corn-growing, but the conditions were better for olives and vines.

In spite of the progress of agriculture and the conversion of pasture-land into arable, stock-breeding, without having the importance which it had once enjoyed, none the less continued to be prosperous. From Asia Minor to Great Greece, from Thessaly and Epeiros to Cyrenaica, the plains and

[1] E. Saglio, in **XVII**, vol. i, pp. 1144 *ff.*
[2] A. Rainaud, in **XVII**, *s.v.* "Silphium."

valleys supported horses, asses, mules, oxen, sheep, and goats. The horses of Thessaly, the asses of Arcadia, the mules of the Peloponnese, the oxen and cows of Epeiros, the sheep of Attica, the Mæander, and the neighbourhood of Taras, and the goats of Scyros were especially celebrated. No farm was so humble that it did not support at least a pig. The poultry-yard contained hens, ducks, geese, guinea-fowl, pigeons. Bees were kept almost everywhere. The honeys of Hymettos in Attica, of the Cyclades, of Cyprus, and of Hyblæa in Sicily were regarded as the best and having the most flavour.

Many regions of the Greek world produced both corn and livestock—for example, Thessaly, Bœotia, Messenia, Elis, the valleys of Asia Minor, Cyrenaica, the neighbourhood of Taras, and that of Leontion in Sicily. Elsewhere the breeding of sheep and, still more, of goats was the chief resource of rugged, mountainous districts, in Ætolia, Acarnania, Arcadia, and most of the islands.

Like the ground devoted to stock-breeding, the wooded areas had been diminished. They were still, however, considerable. The chief mountain masses were clad in deep forest. Macedonia and Thrace furnished much timber for carpentry, and the cities of Great Greece took their supplies from the southern Apennines and the hills of Lucania and Bruttium. Crete, Rhodes, and Lesbos had vast forests. Those of Mysian Ida were famous. The chief species to be found in these woods were the oak, elm, poplar, plane, ash, beech, maple, pine, fir, cypress, yew, and laurel.

These forests, like the copses which abounded in all Greek lands, were not merely stores of wood for fuel and building. They were also hunting-grounds, full of game, and there the Greeks would hunt bears, boars, wolves, foxes, deer, hares, and rabbits, taking them in traps or nets, running them down, or attacking them with large and small spears, javelins, or axes.

Fishing, of which there is little mention in the Homeric and Hesiodic poems, afterwards became important in Greek life. It became the occupation and habitual source of food of many shore-dwelling populations, not only in Greece Proper but on the Gulf of Taras and on the Sicilian coasts in the west and on the Hellespont, Propontis, and Bosphorus in the north-east. Together with vegetable foods and meat furnished by stock-breeding, fish, fresh, dried, or salted, became one of the usual and favourite dishes of the Greeks.

So, from one end of the Hellenic world to the other, land and sea were exploited by the Greeks in the many forms in which men at that time practised agriculture, stock-breeding, hunting, and fishing. Corn, fruit, and vegetables, beasts of land and sea, wild and tame, wood for burning and wood for working, all played an active part in the economic life of the Hellenes.

II

The Equipment and Organization of Farming

Great as was the extension of the ground cultivated and otherwise turned to account by the Greeks, there was no corresponding change or perceptible progress in the equipment or even in the methods of agriculture. Ploughing, sowing, harvest, and threshing, vine-dressing and vintage, are much the same in Xenophon and Theophrastos as in the Homeric poems and Hesiod. The plough was not improved in any way; wheat and barley were still mown with the sickle; the grain was still separated from the ear by the treading of oxen or mules.

But it would be going too far to say that the working of the land had not profited at all by the experience of centuries.

The Greeks strove to adapt their crops to the nature of the soil or to climatic conditions. They distinguished, no doubt by empirical and still superficial methods, the various kinds of soil—fat soil and meagre soil, black soil, sandy soil and clayey soil, dry soil and wet soil. Practice had taught them what crop grew best on such-and-such a soil. So, too, they took into account the rainfall, the wind, the aspect, and the temperature, and knew their influence on vegetation in general and on each of the chief crops in particular.

They saw the imperative necessity of submitting themselves to nature, but they also corrected nature so far as it was possible. They dealt with too steep slopes by terracing them. They made damp, unhealthy plains wholesome by skilful drainage-works, and led water to dry land by irrigation channels. They cleared ground of stones. They altered and improved the quality of certain lands by mixing in other soils in a manner which M. Guiraud compares to the operations which we call marling, liming, and manuring with plaster.[1]

The employment of manures, which had been practised in Homeric times, became more rational. The respective value of various manures, human and animal manure and night-soil, was appreciated, and an attempt was made to determine what sort of manure suited each crop. It is not impossible that the Greeks were acquainted with mineral manures, at least nitre. They also knew that the straw, grass, and vegetable débris left over from certain operations enriched the soil, and that wood-ash, when mixed into the soil of an exhausted piece of land, gave it a little fertility.

The system of fallow was no longer the only method by which the soil was given the rest which it needed. It had been observed that a like result could be obtained by the rotation of crops. The method was still quite rudimentary, and consisted simply in putting down vegetables and cereals in turn. But it was a great advance, all the same, for fallow was at least partly abolished, and it was no longer necessary, for the sake of the future, to leave half the area of arable land idle every year.

So, although in the main there had been no radical change in agricultural equipment and methods in the Greek world, real progress had been made in details. This progress was chiefly inspired by practical experience and daily observation; we cannot yet speak of a science of agriculture. In stock-breeding, no development had taken place. It lost ground in favour of agriculture, but neither its methods nor its organization seems to have been modified.

How was rural property distributed in the Greece of historical times? In our documents we find this distribution varying according to districts and periods. Large estates were the rule in Thessaly and Great Greece, and existed, but less generally, in Bœotia, Attica, and Elis. Medium-sized and small properties prevailed chiefly in the territory of cities with democratic institutions. In Attica there were a few fairly big estates, such as that of Phænippos, estimated at about 740 acres, but most of the landed properties of which we know the value and can roughly reckon the area were of

[1] **XLV**, p. 465.

medium size (from 15 to 125 acres) and far more often quite small (under 15 acres).[1] The land was probably divided into similar small parcels in the hilly parts of Bœotia and those of the Peloponnesian seaboard—Achæa, Elis, and northern Argolis. We have several inscriptions showing that in the islands of the Ægean, such as Tenos and Chios, and in many cities of Asiatic Greece, such as Iasos and Halicarnassos, it was the same.[2] Therefore, even in cases where several of these portions belonged to the same landowner, they were probably exploited as so many small estates.

In Laconia we witness a development in the course of which the big estate supplanted the original small croft among the Spartans. At the beginning, " the soil of Laconia was divided into a great number of portions, the average area of which cannot have been over 17 or 20 acres."[3] According to Plutarch, these allotments were 9,000 in number, and each bore about 80 medimni of corn and fruit (about 165 bushels); this would correspond, so far as such calculations can give any exact result, to an average area of 15 acres. Now, by the time of Aristotle, the land at Sparta was in the hands of a few owners, and about the middle of the third century B.C. their number was down to a hundred. Whatever the legal or social causes of this evolution may have been, the fact is important economically, and its effects on the constitution and political life of Sparta were considerable.[4]

It is true that a development in the same direction took place in the other cities of Greece in the Hellenistic period, but that development was due, among other things, to the conquest of the East, and it is later, in the second part of this work, that we must speak of it. Before the third century, according to the evidence of Aristotle, nobody in Greece was in utter distress, because the land was divided among a great number of holders.[5] In spite of the important place already held by trade and industry in Greek economic life, land still represented the normal source of wealth.[6]

[1] *Ibid.*, pp. 392 *ff.*
[2] *Ibid.*, pp. 396 *ff.*
[3] **XXXVIII**, pp. 99 *ff.*
[4] *Ibid.*, *loc. cit.*; **XLV**, pp. 402 *ff.*
[5] *Pol.*, ii, 3, 6.
[6] **XXXVI**, p. 84.

III

Direct and Indirect Methods of Farming

That wealth was exploited by those who possessed it, either directly or indirectly.

The small crofter cultivated his fields, orchard, and garden himself, with the help of his family and sometimes a slave or two. The " workers of the land " (*georgoi*), so often mentioned and rather chaffed by Aristophanes, were peasant proprietors. The law required the Athenian cleruchs, except in a few very rare cases, to work the land allotted to them in person. At Syracuse, Gelon made every effort to convince his fellow-countrymen who owned land to till it themselves,[1] and Xenophon declares in his *Œconomicus*: " The exercise to which agriculture compels those who cultivate the land with their hands gives them strength."[2]

This direct, personal working of the soil was possible only if the man who owned it lived regularly on his estate and was a practical farmer. Now, many owners of country properties, in Attica and elsewhere, lived in town, and many of them would have made a poor job of driving a plough, handling a scythe or sickle, making wine or oil, and looking after beasts. These had resort to a working staff varying in size, composed chiefly of slaves but also of free day-workers. The supervision of these labourers was entrusted to a bailiff or foreman, usually a slave. There were very large numbers of slaves employed on the land in all Greek cities.[3]

In several regions of Greece, the movements of peoples—invasions, migrations, conquests—which took place at the beginning of the historical period led to a special system of exploiting the soil. Those of the old inhabitants who would not or could not flee before the new-comers were reduced to a condition which is often described as serfdom. This was the fate of the Messenians when conquered by Sparta, and of many peoples on whose land Greek colonies had been founded, for example Pontic Heracleia and Syracuse. The Helots of Laconia and the Penestæ of Thessaly are the best-known instances of these " serfs bound to the soil." The land which they tilled did not belong to them, or rather it no longer

[1] **XLV**, pp. 448 *ff*. [2] v. [3] **XLV**, pp. 452 *ff*.

belonged to them; before the invaders came they had had full ownership of it, and the result of the invasion and conquest was to dispossess them. But they were kept on the soil to till it for the benefit of the conquerors, who now owned it.

Every year they had to deliver a large part of the harvest to the new masters of the land. The latter could not put them to death or sell them out of the country. Relations between landowners and serfs were, moreover, governed by laws or conventions which the masters could not alter; our ancient documents say this definitely of the Penestæ,[1] the Helots,[2] and the Mariandynians of Pontic Heracleia.[3] We do not know exactly how much of the harvest had to be given up by the serfs to the landlords. Plutarch relates that the Helots had to supply, for each lot or *kleros* which they cultivated, 82 medimni of grain and about the same amount of liquid produce (oil and wine), that is, about 165 bushels of each kind. But he does not tell us what proportion of the total harvest these quantities represent. The Messenians, who were conquered and reduced to the same legal and social condition as the Helots, had to furnish the Spartans with half of all the fruits of the earth, and the poet Tyrtæos, who gives us the information, compares them to asses laden with heavy burdens:

Ὥσπερ ὄνοι μεγάλοις ἄχθεσι τειρόμενοι.[4]

We also know that the Helots and Penestæ could acquire movable property, and sometimes much.[5] It therefore seems that the serfs were treated differently in different parts. In any case, serfdom was marked by the same characteristic features in Laconia, in Thessaly, and wherever else it was practised. Of these features, the most definite was the absolute inseparability of the soil and the man who worked it. That union the master could not break by driving or selling the serf away from the lot on which he lived, nor the serf by leaving that lot.

Farming in person, farming by slave and free labourers, and farming by serfs attached to the soil—these were the three

[1] Archemachos, in Athen., vi, 85, p. 264; *F.H.G.*, iv, pp. 314 *ff.*
[2] Plut., *Lycurg.*, 24; Myron of Priene, in Athen., xiv, 74.
[3] Strabo, xii, 3, 4; Athen., vi, 84, p. 264.
[4] Paus., iv., 14, 4-5.
[5] **XXXVIII**, pp. 48 *ff.*, 67.

methods by which the landlord remained in direct contact with the ground which he owned. The Greeks also knew and practised indirect exploitation, by leasing land to tenants in return for a proportion of the harvest (*métayage*) or for a fixed rent.

The system of *métayage* does not seem to have been practised much in Greece. Usually only one instance is cited, and it is still a matter of controversy—the case of the client hectemors (πελάται ἑκτήμοροι) of Attica before the time of Solon. Some regard these as serfs, something like the Helots of Laconia, others as tenants sharing the harvest with the landlord, like the Roman *partiarii*. Does the word "hectemor" mean that they kept only a sixth of the harvest for themselves, or that they handed only a sixth to the landlord? Although Plutarch gives it the latter sense, most modern historians prefer the former interpretation, for the ancient evidence is unanimous in representing the situation of these men as very hard, and this would certainly not have been the case if they had had the use of five-sixths of their crop. M. Guiraud has shown, on the authority of Aristotle, that they were true *métayers*, farming the land on condition that they paid over a portion of the produce. In any case, we hear no more of them after Solon. We do not know how or why the system of *métayage* disappeared.[1]

The leasing of farms for a fixed rent, paid in nature or coin, was very usual in the Hellenic world. There are numerous inscriptions, by which we can determine its methods and conditions. It is true that most of this evidence refers to public or sacred estates, the property of a city or temple, leased to individuals; but we also know of a few such contracts made between private persons, and there seems to be no reason for supposing that the two kinds of agreement were different. The contract of hire was just like a true lease. The term varied; it might be for five years, or for ten years, or even for twenty or forty years. Sometimes a lease was even concluded in perpetuity. Ten years seems to have been a fairly usual term.[2] The rent was paid either in kind or in coin, in measures of corn or in drachmas.

The payment of this rent was not the only condition laid upon the lessee. Many leases specified obligations which

[1] XLV, pp. 421 *ff*.

[2] *Ibid*., p. 425.

certainly had an economic purpose. At Heracleia in Great Greece the tenants of the sacred land of Dionysos were obliged to plant so many vines and so many olives in their portions, were explicitly forbidden to fell, cut, or saw any tree, and had to replace any vines or fruit-trees which should die of old age during the term of their lease.[1] This desire to safeguard and even to develop tree-growing appears in many contracts. We have seen above that the reason for these special stipulations must be sought in the climate of the Mediterranean region. Penalties were provided for any failure to observe the clauses of the lease on the part of the lessee.[2] In all contracts concerning sacred or public domains, it is mentioned that the lessee provides bail, and that this bail must be approved beforehand by the representatives of the city or god owning the estate.[3] The relations between lessor and lessee were, therefore, very exactly laid down. The lessee was not allowed to cultivate the land just as he liked; its future was not left in his hands without reservation or safeguard.

So there were in the Hellenic world, down to about the end of the fourth century B.C., various ways in which landed property was exploited. Some were better suited to the big estate, others to medium-sized and small properties. Some had their origin in extraordinary events, such as invasion, immigration, conquest, or the repression of revolts, while others were the result of agreements between private individuals. In all cases, the object of the Greeks was to secure the best possible treatment for the soil and not, if one may use the expression, to endanger the health and stability of their land by blindly rushing at immediate profits.

IV

The Status and Character of Rural Property in Greece

There is no doubt, and no one denies, that in the fifth and fourth centuries B.C. land in Greece was owned by individuals. Plato says definitely that the law allowed every citizen to dispose of his goods just as he pleased.[4] Aristotle

[1] III, no. xii (pp. 193 *ff.*), tit. ii, §§ 5-7, 10, 13, 20.
[2] *Ibid.*, §§ 9, 11, etc.
[3] *Ibid.*, tit. i, § 2; *cf.* § 4; tit. ii, §§ 17-19.
[4] *Laws*, xi, 923 *ff.*

states that a man's property was divided among his children, whatever their number.[1] This division is ordered in the law of Gortyn, which belongs to the sixth century at the latest.[2]

But it is thought in general that in earlier times the land, at least each of the old allotments (*kleroi*) originally distributed among the citizens of any one city, the κλῆρος πρῶτος or πατρῷος, belonged to the whole family, the *genos*, and not to the head of the family personally; that it remained "the collective property of the whole *genos*."[3] The passages on which this theory is based do not appear to justify such a hard-and-fast conclusion. Some of these passages tell us that in many Greek cities the law forbade the sale of the original allotments; others mention measures or precautions taken to keep the number and size of these allotments the same and so to prevent estates becoming too unequal.[4] Plutarch relates that before the time of Solon the Athenians were not allowed to make wills, the land and house having to remain in the family of the deceased; and that Solon gave every Athenian who had no children the right to dispose of his goods as he pleased, so making those goods the property of those who held them.[5] The same author tells us how and why, at Sparta, the law of Epitadeus gave every father of a family full power to make over his property by deed of gift or will, whereas previously the father had had to leave his *kleros* to his son, whereby the size and number of landed estates had been kept the same.[6]

What do we learn from these various documents? Merely that the right of the father of the family over the early *kleros* was limited by the law, for economic and social reasons, chiefly because the lawgiver wanted to prevent the land from becoming concentrated in too few hands and to save families whose *kleros* had been sold, given, or bequeathed to others than the natural heirs from falling into destitution. What happened at Sparta when the law of Epitadeus was put into force justified those fears. Inequality of fortune became

[1] *Pol.*, ii, 3, 6.

[2] Law of Gortyn, ed. by Dareste, v, 22 *ff.*

[3] **XXXVIII**, p. 35; **XLV**, pp. 53 *ff.*

[4] Aristot., *Pol.*, vi (vii), 2, 5. Sparta: *ibid.*, ii, 6, 10; Heracleides, *Περὶ πολιτειῶν*, ii, 7, in *F.H.G.*, vol. ii, p. 211, 7. Locri: Aristot., *Pol.*, ii, 4, 4.

[5] *Solon*, 21 . . . καὶ τὰ χρήματα κτήματα τῶν ἐχόντων ἐποίησεν.

[6] *Agis and Cleom.*, 5; *cf.* Aristot., *Pol.*, ii, 6, 10.

excessive, for very soon there were a hundred very wealthy citizens as against a multitude who were extremely poor.[1]

It was perhaps to remedy similar consequences that Solon, while allowing the Athenians to dispose of their goods freely, provided that they had no children, made a law limiting the purchase of land.[2] Thereby he tried to prevent the formation of over-large estates. In other cities this limitation of the right of individual ownership over the original *klêros* was inspired by a wish to give people a liking for agriculture, or to prevent them from losing it.

> "To give a people a liking for agriculture," Aristotle says, "there are among the ancient laws of many cities certain most useful dispositions, such as that which forbids the possession of too great an area of land. . . . Originally, the sale of the original *kleros* was in many cities forbidden by law."[3]

Here Aristotle shows clearly, by the manner in which he gives the information, that the whole object of a law of the kind was to protect agriculture. Nowhere is there any question of family or collective property; that is an idea which is not found in the ancient texts and has been introduced into them by modern interpreters.

So, too, the interpretation often given to the case of the *epikleros*, the sole daughter, goes much too far beyond the evidence. The *epikleros* daughter could not herself inherit her father's property. The heir to that property was the son to whom she gave birth, provided that she had married in her father's lifetime, or her future husband had been named by her father in his will, or she had married her father's nearest kinsman, the lawful heir. In any of these three cases, if the inheritance included land, that land sooner or later became the property of the son born of her. After setting forth these rules, Fustel de Coulanges adds: "Rules like these, which doubtless date from a very early time and survived only in fragments in Attic law, are the surest evidence of the principle of family ownership."[4] No. There is nothing in them which derives from that principle. The case of the *epikleros* daughter was governed by quite a different idea—the idea that daughters had no right to the patrimony at all. Will it be said that

[1] Plut., *op. cit.*, 5; *cf.* Aristot., *loc. cit.*, 10 *ff.*
[2] *Ibid.*, ii, 4, 4.
[3] *Pol.*, vi (vii), 2, 5.
[4] **XXXVIII**, p. 42.

landed property in England belongs to the family, because daughters are almost wholly disinherited, chiefly in favour of the eldest son, who gets the landed estate and part of the movable goods ?[1]

That these limitations of the right of individual ownership had economic effects, no one will deny. The circulation of land, as a merchandise, was slowed down by them. For a long time, the small and the medium-sized property were better able to hold their own against the big estate. Rural families remained attached to the soil, since the original allotments which formed the core of their estates could not be sold or given or bequeathed outside the *genos*. A philosopher —Plato—might say that in his view a property belonged, not to the individual who held it, but to his whole family, dead and unborn; but that is the expression of a moral idea, and does not at all imply that the notion of family ownership prevailed. So, too, when Plato adds that every family with its goods belongs to the State, we cannot conclude that in Greece, even in primitive Greece, the property-system was communistic.

In the Greek world, land was everywhere owned by individuals. There is not a text which authorizes us to suppose that a piece of land belonging to the head of a family was regarded as belonging at the same time to the other members of the family. All that we know is that the right of property was not unrestricted in earlier times. The making over, whatever form it might take, of the original lots could not be effected at complete liberty, without any guarantee for the economic and social equilibrium of the citizen-body. What, in the eyes of the ancient philosophers, justifies these limitations of individual right is the public good, the interest of the city, not concern for the family. The object of the lawgivers who made these restrictions seems to have been a somewhat chimerical equality of wealth. Aristotle has no difficulty in showing the futility of such a policy. At all events, the notion of family ownership, from the legal or from the economic point of view, does not appear in any of our documents.

[1] Boutmy, *Psychologie politique du peuple anglais*, pp. 300 *ff*.

CHAPTER IV

INDUSTRY AND INDUSTRIAL LABOUR IN THE GREEK WORLD FROM THE SIXTH TO THE FOURTH CENTURY BEFORE CHRIST

THE expansion of the Greeks over the Mediterranean and the progress of Greek colonization had no less effect on the development of industry and industrial labour than on that of agriculture.

The commercial relations which steadily increased between the Greek cities (of the mother-country and of the colonized regions) and foreign nations could not consist solely in the importation by the Greeks of the foodstuffs and other products which they lacked. These importations had to be met by exportations. But Greece had no foodstuffs or raw materials for this business; manufactured goods were all that she could offer.[1] Industry became an absolute necessity to her economic welfare. In early times, industry had had to meet only domestic and local needs; henceforth it would have to provide commerce with indispensable articles of exchange.

The change was favoured by many causes. Formerly the Phœnicians had been, for the Greeks themselves and for the other peoples of the Mediterranean, the sole providers of most of the industrial products which could not be made in the family. It was they who had brought the many creations of the arts and industries of the East to the beaches of the Ægean and Ionian Seas. When they had been driven from those seas and markets by the Greeks, the Hellenic industries found outlets and customers there, and to take advantage of the former and to satisfy the latter they were compelled to develop. So, between industry and trade (chiefly sea-borne) action and reaction took place, trade needing to be fed by industry and industry owing its prosperity to trade.

[1] This seems to have been the idea which led Solon to make laws encouraging industry in Athens, and particularly the artistic industries (Plut., *Solon*, 22).

Secondly, the Greeks, whose fleets had become predominant in the Eastern and Central Mediterranean, came into immediate contact with the countries which produced or sold raw materials—with the various points on the coast of the Euxine, where the copper and iron routes ended; with Cyprus, the island of copper; with Thrace, rich in precious metals and timber; with the places on the coast of Asia Minor, to which the wool of Phrygia and Lydia came down; with southern Gaul, where one of the roads taken by the tin ended; and with Egypt and Libya, which supplied ivory.

In Greece itself, certain industries which found all the raw materials which they needed on the spot, like the ceramic industry, developed more or less spontaneously, together with the metallurgical and textile industries, to which the new economic conditions had given a lively impulse.

Everything, even the accidents of political life, contributed to the progress of industry in the Greek world. It cannot be denied that the industrial prosperity of cities like Miletos, Athens, Corinth, and Syracuse was greatly influenced by the incidents of their domestic and foreign politics. To mention only one instance, the victory of Themistocles over Aristeides and the ostracism of the latter in 483 gave an impetus to many Athenian industries.

The result of these different causes and conditions was that industry made great progress in the Greek world in the fifth and fourth centuries. It was especially active in Attica, in Athens and at the Piræus. In his *L'Industrie dans la Grèce ancienne*, M. Francotte calls the sixth, fifth, and fourth centuries the "Athenian period,"[1] and so, too, M. Glotz, in his *Ancient Greece at Work*, places the "Athenian period" between the "Archaic period" (ending about the middle of the sixth century) and the "Hellenistic period."[2] This does not mean that all industrial activity was at that time concentrated at the foot of the Acropolis and on the wharves of the neighbouring port. It was going on in other places, in Greece itself and in various colonies. But it was nowhere so intense, and it nowhere had such a place in the economy of the State, as in Athens.

[1] XLII, p. 38. [2] XLIV.

I

Raw Materials. Their Sources and Employment

Of the raw materials transformed by Greek industry, some were to be found in Greece itself or in the regions colonized by the Greeks, while others were procured from abroad.

The oil-works everywhere could obtain a sufficient supply of olives. The flour-mills were not in the same case. Greece Proper was poor in cereals, and Sicily and southern Italy had not enough to cover the deficit in the national production. The Greeks had to draw on Egypt and the plains of Scythia, and from those two countries huge cargoes of corn were every year unloaded at the Piræus.

There was no lack of building-materials in Greece. There were many stone-quarries; we hear in particular of those of Thessaly, Attica, Argolis, and Laconia. Some yielded calcareous tufa, which was much used for buildings. Others, especially in Laconia, supplied a green porphyry which was in great demand. We know more about the marble-quarries. In addition to the famous marbles of Pentelicon and Paros, the Greeks worked quarries west of Larissa in Thessaly, on Hymettos in Attica, in several ranges in the Peloponnese, for example in Parnon and on Cape Tænaron, on the island of Naxos, on Proconnesos in the Propontis, and in the environs of Ephesos.[1] Clay was found almost everywhere.

The forests, which were still numerous, especially in Macedonia, Thessaly, Arcadia, Crete, and the Troad, furnished various timbers—pine, fir, oak, poplar, elm, ash, maple, cedar, cypress, beech, and plane. However, the Greeks went to Asia for some—to Cilicia and Syria for cedar, to the country about Pontos for ash, and to Phrygia for plane.[2]

Of the textile materials, wool was the only one which Greek lands could supply in abundance. From this point of view, the most important districts in Greece Proper were Attica and Arcadia, the wool of Attica having a special reputation. It could also be obtained in Thessaly, Bœotia, the Megarid, the islands of Eubœa, Samos, Cos, and Cyprus, and

[1] A. Jacob, in **XVII**, *s.v.* "Lapides"; G. Lafaye, *ibid.*, *s.v.* "Marmor."
[2] A. Jacob, *ibid.*, *s.v.* "Ligna."

Great Greece. Native production did not suffice, and wool had to be imported from Lydia, Phrygia, and countries round the Euxine. Miletos was the principal market.[1] Flax was grown in Macedonia and Thrace, but most of what was woven in Greece came from the East, from Colchis, Asia Minor, and Egypt.[2] The plant called byssus by the ancients, which has sometimes been taken for cotton but seems rather to be like flax, also came from the East. It was grown in Elis, but the fact is only mentioned by Pausanias, and we do not know how early it was introduced into Greece.[3] Hemp, which was used for making ropes, does not seem to have been cultivated in Greek lands; according to Herodotos, there was plenty of it among the Scythians, and the Greeks probably obtained their supplies in the colonies on the northern coast of the Euxine.[4]

The oxen, sheep, and goats reared by the Greeks themselves gave them the various kinds of leather needed for the tanning, leather-dressing, and shoemaking industries,[5] but furry animals, although not rare in certain regions such as Bœotia and Arcadia, were not sufficiently numerous to supply the necessities and luxuries of the Greek townsmen; furs from Sarmatia and Libya were greatly prized in Greece.[6]

Although not completely destitute of metals and metallic ores, the Greek world was not favoured in this respect. According to many traditions, gold came from distant regions—from the land of the Arimaspians (perhaps the Ural), from that of the Argippæi and Issedones, which some place in southern Siberia, from Colchis, and from Asia Minor, particularly Lydia. In Greece itself, a few veins were worked in the islands of Siphnos and Thasos, in Macedonia, and on Mount Pangæos in Thrace.[7]

With silver it was the same as with gold. First the Greeks obtained this metal from the East, perhaps from the country of the Chalybes (north of Armenia) and Colchis, and then they found it on their own soil and were able to extract it. Here silver-working was closely connected with lead-working; the famous mines of the Laureion in Attica yielded

[1] H. Thédenat, *ibid.*, *s.v.* "Lana."
[2] Id., *ibid.*, *s.v.* "Linum."
[3] G. Paris, *ibid.*, *s.v.* "Byssus."
[4] **XLV**, p. 501.
[5] G. Lafaye, in **XVII**, *s.v.* "Corium."
[6] M. Bernier, *ibid.*, *s.v.* "Pelles."
[7] L. de Ronchaud, *ibid.*, *s.v.* "Aurum."

argentiferous lead, and the same is probably true of Mount Pangæos and Siphnos, where silver was obtained.[1]

True lead-mines were more numerous, and yielded, in addition to the metal, a variety of products such as red-lead, ochre, litharge, etc. The ancients tell us of lead-veins in Ceos, Seriphos, Anaphe, Rhodes, Cyprus, and Chalcidice.[2]

Copper was at first imported among the Greeks by the Phœnicians; then the Greeks went for it themselves to Cyprus. They found deposits in Eubœa, near Chalcis and Ædepsos, and it has been supposed that it was to be found in every part of Greece where a city had the name of Chalcis, "Bronze Town."[3]

Tin, which is needed for the manufacture of bronze, does not exist in Greece nor in any eastern country near it. It was entirely imported from the western parts of Europe—the Peninsula, Gaul, and Britain. Where exactly were the famous Cassiterides, the Tin Islands (*kassiteros*=tin)? They are sometimes identified with the Scilly Isles, at the south-western end of England. But the problem is far from being settled. According to a recent theory, they were the islands about Quiberon Bay, off the west coast of France. The metal was first brought to Greece by the Phœnicians, and later by the Carthaginians, who went both direct through Gaul and by sea to the countries which produced it. The Greeks subsequently emancipated themselves from this dependence. The tin was carried overland through Gaul to Marseilles, and there Greek ships took it on board for the Hellenic countries.[4]

There was not much iron in the soil of Greece. The ancient authors speak of it, or traces of workings have been found, on the peninsula of Tænaron in Laconia, in Bœotia, in Eubœa, and in several islands of the Ægean—Andros, Syros, Seriphos, and Scyros. According to the legends, the Greeks obtained iron and learned how to work and use it from various peoples of Asia Minor, especially the Phrygians and Chalybes.[5]

We should also mention among the raw materials used by Greek industry ivory and amber. Before the expedition of Alexander, which opened the routes to India to the Greeks,

[1] E. Saglio, *ibid.*, *s.v.* "Argentum."
[2] M. Besnier, *ibid.*, *s.v.* "Plumbum."
[3] W. Cart, *ibid.*, *s.v.* "Æs."
[4] M. Besnier, *ibid.*, *s.v.* "Stannum."
[5] L. de Launay, *ibid.*, *s.v.* "Ferrum."

the ivory used in the Greek world was of African origin. It is said to have come from Ethiopia and from Western Africa, that is, from the present French colony of Mauritanie (N. of Senegal).[1] The amber was said to come from the North. The two roads by which it came to the Mediterranean ended on the north coast of the Euxine and at the head of the Adriatic respectively; that is why tradition sometimes reported that it came from Scythia and sometimes that it was found round the mouths of the Po.[2]

Greece, then, was far from being self-sufficient in respect of the raw materials required for industry. These natural conditions helped to encourage a twofold current of trade—importation to bring those materials to Greece and exportation to take manufactures abroad, in order to establish what is today called the balance of trade. Another logical and inevitable consequence of the situation was that Greek industry had to produce more than was required for local consumption; it was obliged to feed a fairly big export trade.

II

The General Character of Industry. The Division of Labour and Specialization

In the society of the Homeric and Hesiodic poems, industry was mainly domestic.[3] With a very few exceptions, all that was needed for daily life was made in the house, and what household labour, free or slave, could not supply was usually bought from foreign traders, the Phœnicians. In exchange for the fine stuffs of the East, embroidered or dyed purple, and tripods, bowls, and vases of precious metal, the Greeks of that age gave slaves or livestock. The only specialized crafts mentioned are those of the smith, the armourer, the builder, and the potter.[4]

Another characteristic of this household industry was that one man could do very different kinds of work. Odysseus was a boat-builder and cabinet-maker as well as a ploughman and reaper. Eumæos cut leather and made his shoes, just as he built the styes for his herds of swine. In the *Works and*

[1] A. Jacob, *ibid.*, *s.v.* "Ebur."
[2] Id., *ibid.*, *s.v.* "Electrum."
[3] See above, p. 18.
[4] See above, p. 18.

Days, Hesiod shows us the small landowner making his own tools, plough, and cart. The various operations of the textile and food industries were done by the women in each house.

In the classical period, from the sixth to the fourth century, the spectacle offered by the Greek world at work is quite different. No doubt, many crafts were still practised in every home. Not only in the country, but in the city, the mother of the family made the bread, or at least supervised the process, and gave out wool to her daughters and slaves to spin. There were still Greek peasants who did not call in a craftsman to make or repair their implements. But the characteristic of this new period was that the professions had become more numerous, organized, and specialized. The division of labour had been greatly developed. In the speech in which he defended his administration, Pericles mentioned many industries to which his policy had given an impetus: he spoke of carpenters, sculptors or modellers, bronze-workers, stone-cutters, gold-beaters, ivory-turners, painters, embroiderers, metal-chasers, cartwrights, rope-makers, linen-weavers, paviours, cobblers, and miners.[1] Xenophon says that in his time, in the big towns, each craftsman practised only one trade, and sometimes they were even more specialized; there were shoemakers who made only men's shoes and others who made only women's, one could not or would not sew the shoes, another was incapable of cutting the leather, and so on.[2] Plato, in the *Laws*, recommends that no man who works in iron should at the same time work in wood, stating that there is practically no one who can excel in two different crafts at once.[3] In the *Republic* he expresses the same idea, when he makes Socrates utter sentiments like the following:

> "More things are made, and they are made better and more easily, when each man does the work to which he is suited"; or, again: "The ploughman must not himself make his plough for his own use, if he wants it to be well made, nor his mattock, nor the other implements used for tilling the soil. It is the same with the architect, who needs many tools, and with the weaver and with the shoemaker."[4]

A rapid survey of the principal industries then existing in the Greek world will better show up this new character of industrial labour.

[1] Plut., *Pericl.*, 12.
[2] *Cyrop.*, viii, 2, 5.
[3] viii, 846D-E.
[4] ii, 5, 1 (370).

CHAPTER V

TRADE IN THE GREEK WORLD FROM THE SIXTH TO THE FOURTH CENTURY BEFORE CHRIST

THE new conditions of farming and the rise of industry transformed the economic life of the Greek world more and more. In Homeric and Hesiodic society, that life is chiefly of a domestic, family nature. Like Alcinoos and Odysseus, Perses obtains most of the food which he needs on his own land and himself makes his tools and implements; purchases outside the household are quite exceptional, and the little trade which is done takes the form of barter.

The expansion of the Greeks over the Eastern and Central Mediterranean, the foundation and growth of many prosperous cities, the birth and development of increasingly specialized industries, the ever more marked divergence between production and consumption—all these facts produced an economic evolution which reached its height in the fifth century and in the first two-thirds of the fourth. The effects of that evolution did not reach beyond the confines of the Hellenic world, but within those limits they in their turn exercised a mighty influence on the social and political life of Greece. Like life in general, economic activity became urban, and extended to relations between cities and with foreign peoples and states—the Barbarians, as the Greeks called them. This new economic life was subject to necessities not known before. It imposed duties on the cities in which it developed, but it brought them great advantages. By the side of the landed wealth, agricultural and pastoral, of past ages it produced movable wealth, which was more fluid, more varied, and perhaps more living and fruitful.

The principal element in that economic life, by the side of industry and partly due to it, was trade in all its forms, with its manifold aspects, its already complex organization, its equipment, and its laws and regulations.

I

The Nature of Exchanges. The Principal Kinds of Merchandise

The simplest form of trade is that which is conducted mainly in the market of each town or human agglomeration, to which one should add local shops, which are usually mere booths open to every wind that blows. That is the meeting-place of the humble producers who sell their farm-produce or handiwork direct and the retailers or pedlars who are already middlemen between producer and consumer. The Agora of Athens presented a most animated spectacle. "The market-gardener came to town with his fruit and his vegetables, the landowner sent in his asses laden with wood, the Acharnian brought his baskets of charcoal. . . . All who have something to sell, slaves with cloth which they have just made, craftsmen from the Cerameicos, Melite, or Scambonidæ, peasants who left their village before daybreak, Megarians driving their pigs, fishermen from Lake Copaïs, pass in every direction. Through alleys planted with trees they reach the places assigned to different goods, separated by movable barriers. One after another, at the hours fixed by the regulations, the different markets open; there are markets for vegetables, fruit, cheese, fish, meat and sausages, poultry and game, wine, wood, pottery, ironmongery, and old articles. There is even a corner for books."[1] Special magistrates, the *Agoranomoi*, keep order, inspect and control the holding of the market, and decide any disputes which may arise between merchants and customers. Some cities had covered markets, either devoted to one form of trade, such as the corn-market, or open to all merchants. Gradually covered markets took the place of open-air ones in the Greek cities generally.[2]

These markets, covered or otherwise, satisfied local needs; both represented the purely urban element of trade. With the great fairs, periodical and other, the economic framework began to extend. The most important of these fairs were those connected with certain religious feasts, the national games at Olympia, Delphi, Nemea, and the Isthmus, the

[1] **XLIV**, English pp. 288 *ff*.
[2] P. Huvelin, in **XVII**, *s.v.* "Mercator," p. 1735.

Panathenæa in Athens, the Delia at Delos in honour of Apollo, etc.[1] The sacred truces which were declared in order to give all security to the pilgrims who came pouring from the various parts of the Greek world were used by merchants no less than by worshippers. Armies in the field were followed by swarms of traders, who procured for the troops all that they needed and constituted veritable moving fairs. Thucydides relates that the Sicilian Expedition comprised, in addition to State cargo-boats and transports, many vessels sailing on their own account, to supply the markets.[2] But this was not a permanent traffic. The markets in question were determined by special or exceptional circumstances, whether they themselves were fixed or moving, and had only a small place in the commercial activity of the Greek world.

Much more important were the exchanges which went on between Greek cities, or between Greeks and barbarians. It is not easy to distinguish between the two classes of trade. Miletos, Cyrene, and the colonies of the Euxine did not export only the products of their own soil; they were also shipping centres, situated as it were on the circumference of the Greek world, and in that capacity they received many foodstuffs and raw materials from foreign countries and passed them on to other Greek cities, and also sold the articles manufactured by Greek artists and craftsmen among the barbarians. The current of trade between Greek lands and cities was not fed only by Greek goods. With the silphium and pottery produced in her own territory, Cyrene sent to Greece goods brought to her market-place by caravans from the centre of Africa. In addition to the salt fish from their own fisheries, the colonies of the Euxine supplied many Greek cities with corn grown by the Scythians and amber from the country of the Hyperboreians. Miletos was not only an industrial centre, exporting its local products, but an intermediary between the Asiatic East and Greece Proper. Complex in source and character, the trade between cities and between nations which had grown up in the Greek world as a consequence of the colonial expansion of the Greeks dealt chiefly with foodstuffs, various raw materials, and all kinds of manufactured

[1] Id., *ibid.*, *s.v.* "Mercatura," p. 1765; *cf.* Laloux and Monceaux, *Restauration d'Olympie*, p. 182; Defrasse and Lechat, *Épidaure*, p. 234.

[2] vi, 44.

articles, mainly produced by the textile, ceramic, and metal industries.

Wheat, wine, oil, certain fruit like figs and almonds, and salt fish went from one end of the Greek Mediterranean to the other, from southern Italy and Sicily to the coasts of the Crimea and Armenia, from Cyprus and the Nile Delta to the top of the Adriatic. Metals, useful or precious, wood, wool, hides, amber, and ivory converged on the industrial centres, especially Athens and Corinth. Woven materials, coarse and fine, carpets, painted vases, weapons of all kinds, metal utensils, jewels, works of art, and the products of the goldsmith and metal-chaser were carried by ships over the seas in every direction.[1] If we are to reach a fair estimate of this economic movement, we must be careful not to exaggerate any more than to understate. Obviously, it cannot be compared to the commercial activity of modern times. Any analogy between the ports of ancient Greece and modern Genoa or Marseilles will provoke only scepticism or a smile. Nevertheless, the spectacle afforded by all this exchange, shipping, and coming and going of goods was then a new thing in the Mediterranean. It was quite different, in intensity and in nature, from that previously afforded by Phœnician trade, which had been mere sea-peddling rather than real business.

II

Trade-Routes by Land and Sea. Chief Ports and Commercial Centres. The Commercial Expansion of the Greeks

The great trade-route of the Greeks was always the sea. The land-roads were not, however, neglected. Two in particular seem to have been used by traders. Both started from Attica, one running over the Isthmus of Corinth to the Peloponnese, where it passed through Argolis and eastern Arcadia to the valley of the Eurotas, with a branch through Argolis and northern Arcadia to Elis and Olympia, and the other going northwards and north-eastwards, passing through Bœotia to Delphi and, perhaps, through Phocis to Thessaly.

[1] **XLII**, pp. 132 *ff.*; **XLIV**, English pp. 118 *ff.*, 310 *ff.*; P. Huvelin, in **XVII**, *s.v.* "Mercatura," pp. 1763 *ff.*

These were pilgrims' ways rather than trade-routes, but we may suppose that merchants took them as well as worshippers. As for the routes of Asia Minor, the caravan-roads which ended in Cyrenaica, and the two ways by which the amber came across Europe from the Baltic to the Black Sea and the Adriatic, they were outside the Hellenic world, and we cannot regard them as characteristic elements of Greek commerce.

The sea, on the other hand, played the most active part in the commercial life of the Hellenes. By the sea alone did Greece Proper communicate with its colonies, whether the larger territories of Asiatic Ionia, Cyrenaica, and Great Greece, or the settlements and trading-stations scattered along the coasts of the Euxine, on the Bosphorus and Hellespont, on the seaboard of Thrace and Macedonia, in Cyprus, in the Delta of the Nile, and in the Far West among the Ligurians, Gauls, and Iberians. In the fifth century, the Piræus had become the centre of a number of convergent sea-routes, which one might almost call regular lines of shipping. To the north and north-east, in addition to the coastal route which served the colonies in Macedonia, Chalcidice, and Thrace, there was the great route to the Hellespont and the Euxine, which was of capital importance to the Athenians, since they largely depended on it for their supplies of corn and dried fish. Eastwards, over the Ægean and through the Cyclades, ships sailing from the Piræus reached the islands and principal ports of Asia Minor—Lesbos, Chios, Samos, Phocæa, Smyrna, Ephesos, and Miletos. Towards the south-east, the Greeks went out of the Ægean between Crete and Rhodes, and reached Cyprus, the Phœnician ports, and the busy and prosperous trading-station of Naucratis. There was equal opportunity for their commercial expansion in the Ionian Sea and the Central Mediterranean. After rounding the southern points of the Peloponnese, the navigator might make due south for Cyrenaica, west for Sicily, or north-west for Great Greece and the Adriatic, up which sea he might go as far as Hadria and the mouths of the Po. Beyond Great Greece, Marseilles and its neighbours, dotted along the coast from Nice to Rosas, were the furthest outposts of Greek trade in the West.

Marine activity of this kind could not be explained if the

art of navigation and the material organization of the ports had not reached a certain development.

The old boats described in the Homeric poems and the vessels used for the seasonal coasting-trade of which Hesiod speaks had been left far behind. To navigate the Mediterranean, the Greeks could now build ships in which the two driving-powers then known, wind and rowing, could be used at once or alternately. Cargo-boats carried a very large stretch of canvas, supported by a strong mast and yard. The number of banks of rowers was increased; the trireme dates from the end of the eighth century. As a rule, merchant vessels used their sails, oars being used only in exceptional circumstances. They could carry up to 250 tons of goods; regarding their tonnage properly so called, that is, their capacity, we have no exact data. The fact that the sail was most generally used made navigation subject to the wind-system, principally in the Ægean. In spring, the wind blew from the south, bearing the ships from Attica to Thrace, the Hellespont, and the Euxine. At the end of May, the Etesian winds began to blow from the north and north-east, and went on till September, bringing the ships back from those regions before the end of the autumn. The Greeks were not afraid of the high seas, but they did not lose their habit of hugging the coasts as much as possible and keeping land in sight. Many, to reach Sicily, went right round the Ionian Sea, by Corcyra, the Iapygian Promontory, and the coast of Bruttium. In any case, whatever their habits and traditions, Greek vessels ploughed the seas from the slopes of the Caucasus to the Pyrenæan headlands and from the swamps of the Po to the sands and rocks of Cyrenaica.[1]

A big merchant navy of fairly large vessels needed specially equipped harbours. The old beaches, on which boats were drawn up high and dry, were no longer sufficient. Basins were dug and jetties, breakwaters, and wharves were built to protect ships riding at anchor from the swell of the open sea and to facilitate the unloading of cargo. The Piræus, the ports of Corinth, Mitylene, Samos, Ephesos, Miletos, Cnidos, Syracuse, and other harbours yet, bore witness to the technical competence of the engineers. On their sites one may still

[1] **XLIV,** English pp. 115-16, 293 *ff.*; P. Huvelin, *art. cit.*, pp. 1766-67; C. Torr, in **XVI,** *s.v.* "Navis."

see many vestiges of the constructions to which they owed much of their prosperity. Behind the harbour stood warehouses, to receive the goods unloaded from the ships, and there were big squares and wide streets for traffic.[1]

There were many ports on all the coasts of the Mediterranean occupied by Greeks. Some of them played an especially active part in Greek economic affairs. Towards the middle and through the second half of the fifth century, the Piræus eclipsed the other marine centres of Greece, but, before those great days of the Attic port, Ægina and Megara on the Saronic Gulf, the two ports of Corinth, one facing the Ægean and the other the Ionian Sea, and Chalcis on the Euripos had been the great rivals in trade and sea traffic. Delos in the Cyclades, Samos, Chios, and Lesbos nearer the Asiatic coast, and Miletos and Phocæa in Asia itself were centres of intense economic life. On the route to the Euxine, the ports of the Hellespont, Propontis, and Bosphorus, among others Abydos, Cyzicos, and Byzantion, offered well-protected berths to ships calling in. Sinope and Trapezus, the outlets of Armenia, Dioscurias in Colchis, and Panticapæon, Theodosia, and Olbia, the markets at the ends of the roads from Scythia, the Hyperboreian country, and the distant depths of Asia, received the products of the Greek industries and exchanged them against corn, fish, raw materials, gold, and amber. On the Eastern Mediterranean, the ports of Cyprus, the station of Naucratis on an arm of the Nile, and Cyrene were, in different forms and in different conditions, centres of traffic and points of contact between Greeks on the one hand and Syrians, Egyptians, and Libyans on the other.

In the west, Corcyra showed the way and served as a port of call to ships entering the Adriatic. Taras and Syracuse on the Ionian Sea and Cyme on the Tyrrhenian surpassed all other cities of Great Greece and Sicily in prosperity, and it was due, at least in part, to their commercial connexions. Lastly, on the north of the Western Mediterranean, Marseilles took in and distributed abundant valuable goods, from Gaul to the Greek and Eastern world and from that world to Gaul.

The commercial activity of the Greeks went beyond the regions where they had founded colonies. What were the furthest points reached by it, directly or indirectly? In Asia

[1] M. Besnier, in **XVII**, *s.v.* "Portus," pp. 596 *ff.*

and towards the East, Greek goods do not seem to have penetrated very far. No doubt, before the Persian Wars, there was much buying and selling of foodstuffs, raw materials, and manufactured goods between the Greeks and the Lydians and Phœnicians, and after the great conflict, during which the Persian Empire naturally closed its doors to Greek traders, intercourse was resumed. Greece certainly exported some products of its industries to the East and sent some merchants to visit its recent enemies. Naucratis, too, in the hands of the Greeks was a very busy entrepôt and acted as a point of contact between Greece and Egypt. But these economic relations have left no traces to be compared with what we now know of the commercial penetration of the Hellenes into the West and into Central Europe.

In North Africa, Carthage, although she was their rival and often their enemy, kept up an active business with the Greek cities of Sicily. In the Punic cemeteries, excavators have found Greek vases, chiefly Proto-Corinthian and Corinthian, terra-cotta statuettes, bronzes, and engraved and sculptured ivory and bone, the Greek origin of which is undeniable. Syracuse was probably the market where the Carthaginians obtained such articles.[1] According to the *Periplus* of Scylax, the Phœnicians, who went to the island of Cerne, opposite the Canaries, to do trade with the west coast of Africa, used to take Attic vases there.[2]

For Western and Central Europe, M. Déchelette has made a list and a map of all sites north of the Alps at which Greek, Italo-Greek, or Etruscan objects have been found.[3] The extreme points on this map, working round from the west by the north to the north-east, are Clermont-Dessous, in the French Department of Lot-et-Garonne (fragment of a black-figure Attic vase), Clermont-Ferrand and environs (bronze œnochoê-handle), Bourges and environs (several bronze œnochoæ), St.-Jean Trolimon in Finistère (fragments of a bronze helmet), Mook in Holland (bronze œnochoë), Bosdorf in Oberhessen (bronze hydria-handle representing two wrestlers), Langaa in Denmark (bronze vase), Keldby in Denmark (bronze situla adorned with palmettes), and Vettersfelde Guben in Brandenburg (fish and ornaments of repoussé gold). On the

[1] **LXXI,** vol. iv, pp. 154 *ff.*, 162 *ff.* [2] *Ibid.*, p. 142.
[3] **XIX,** vol. ii, pt. iii, pp. 1595 *ff.*, map v.

tin-route, the Greeks went at least as far as the Isle of Wight. "The tin collected in the mines was smelted, purified, and made into ingots, which were conveyed by land to a point opposite the Isle of Wight. On that island it was sold and handed over to the merchants from the south, who, sheltered from storms, took it on board, and carried it thence to the rivers of Gaul."[1] So, too, in the Iberian Peninsula, the Greeks doubtless tried to reach the metal-deposits of the Cantabrian country.[2]

In the vast steppes extending northwards from the Euxine, all along the gold and amber routes, towards the Ural and the Baltic, objects of gold and silver made their way, and many beautiful specimens have been found in tombs in southern Russia—crowns, diadems, earrings, pendants, vases, etc.

Inside the huge area contained within these extreme points, two regions felt the influence of the Greeks more than others—the Gallic or Ligurian seaboard of the Mediterranean, where the finds of Greek pottery at Montlaurès and Enserune and in the environs of Marseilles have proved the diffusion of goods from Corinth and Athens, and Etruria, the tombs of which have yielded up so many masterpieces of the potters and goldsmiths of Greece.

Thus the commercial activity of the Greek world affected many barbarian countries. In the search for raw materials and markets, Greek industry and art made their way along distant trails, to the English Channel, the North Sea, the Baltic, and the unknown regions where Europe meets Asia. Before that great event, the expedition of Alexander, before all barriers between Greece and the East were thrown down, the enterprising genius of the Greeks had penetrated west and north far beyond the points where it had succeeded in founding colonies and trading-stations.

III

MONEY. THE ORGANIZATION OF EXCHANGES

For trade of this kind to be able to develop successfully over such a vast area, more highly developed methods of exchange were needed than the simple barter of Homeric and

[1] **XCVI**, vol. i, p. 410. [2] *Ibid.*, pp. 412-13.

Hesiodic times. The invention of money and the spread of its use must be reckoned among the most potent causes of that economic activity. At an early date, certain metals provided an instrument of exchange which was much appreciated, in the form either of ingots or of instruments, utensils, and weapons, such as cauldrons, tripods, axes, and spits of metal. For this purpose bronze and copper were in especial request. But in practice these masses of metal, raw or manufactured, were awkward to use. The really fruitful revolution, which gave trade an impulse which it had not known before, was the invention of true money, coin with a public, official stamp. " The complete State coin," M. Babelon says, " is an ingot of precious metal stamped with the official mark of the public authority which guarantees its weight and alloy." Or, again, " A coin is only an ingot weighed and guaranteed by the State."[1]

The earliest coins now known date from the seventh century. Some of them were struck by various Greek colonies in Asia Minor, such as Miletos, Phocæa, and Ephesos, others in Argolis, Ægina, Eubœa, and Attica.[2] These coins were made of precious metal, gold, silver, or electrum (pale gold).[3] They represented values which were big for the time in a small compass. M. Glotz has reckoned that in the time of Pericles a single man could live comfortably—that is obtain all he needed and doubtless some superfluities—on 120 Attic drachmas a year.[4] Now, the coin in commonest use in Athens was the tetradrachm, so that 120 drachmas took the material form of thirty silver coins, or, as they were called, thirty " owls," in allusion to the image on the reverse.

For us, who have so long been accustomed to handling coins, it is hard to realize the enormous advance represented by the use of these pieces of gold, silver, and electrum as instruments of exchange. Speaking of the diffusion of the coins of Phocæa and Marseilles in Gaul, M. Camille Jullian writes: " Light, shining, adorned with curious new emblems and a variety of vigorous, striking images, these little disks of metal, to which the Greeks clung with respect as to fetishes

[1] **XL**, pp. 9, 13. [2] *Ibid.*, pp. 13 *ff*.

[3] This electrum, the pale gold of the ancients, was an alloy, natural or artificial, of gold and silver.

[4] **XLIV**, English p. 286.

of wealth, must have made a deep impression on the barbarians."[1] One may be allowed to suppose that the same impression was felt by the Greeks themselves, in Hellas and in Asia Minor, when those same disks of metal first appeared in the place of tripods, cauldrons, axes, and spits of metal for the settlement of purchases and sales.

This does not mean that this monetary economy was simple and convenient from the very first. Every city struck its own coinage, and monetary systems differed, if not between city and city, at least between one region and another. Moreover, coins were made of three metals of unequal value, gold, electrum, and silver.

However, the disadvantages inherent in this variety do not seem to have greatly hampered commercial transactions. M. Glotz speaks of a hoard of six hundred coins found at Taras, " coming not only from every town in Italy and Sicily, but from Corinth and Corcyra, the Cyclades and Cyrene, Athens and Ægina, Thrace and Macedonia, Chios and Phocæa."[2] These coins belonged to different systems, and yet the people of Taras managed to find them all useful.

The calculations needed to co-ordinate the respective values of coins were not very complicated. Everybody knew that in the Æginetan system, which prevailed all over the Peloponnese, the silver drachma or half-stater weighed 6·28 grammes, that in the Euboic or Euboico-Attic system the same coin weighed 4·36 grammes, and that in the Milesian system the silver standard was a stater of 14·52 grammes. He also knew that, whereas in European Greece only silver was minted, in Asia Minor gold, electrum, and silver were all coined simultaneously, and that the relative values of the three metals were as follows: gold was to silver as 1 to $13\frac{1}{3}$ and to electrum as 1 to $1\frac{1}{3}$, and electrum was to silver as 1 to 10.[3] The Greeks had been used to handling the coins struck in Lydia since the reign of Crœsus, and in the Persian Empire since the time of Darius. In Lydia, the gold stater weighed 8·17 grammes and the silver stater 10·89 grammes; in Persia, the gold daric weighed 8·41 grammes and the silver daric 11·21 grammes. Operations of exchange between the principal types of coinage were facilitated by the relationship

[1] **XCVI**, vol. i, p. 223. [2] **XLIV**, English p. 122.
[3] **XL**, p. 19.

existing between subdivisions of the various systems. Thus, in the Æginetan system the four-obol piece, weighing 4·25 grammes, was taken as equivalent to the Attic drachma of 4·36 grammes, and, again, the Corinthian drachma corresponded to the Attic four-obol piece.[1]

Another form of monetary agreement and co-operation was the conclusion of veritable unions between two or more cities—for example, between Phocæa and Mitylene, between Corinth and her western colonies, between all the cities of the Bœotian League,[2] and between the Achæan colonies of Great Greece.[3] During the fifth century, the political hegemony of Athens and the formation of her sea-empire resulted in the predominance of the Attic coinage, and particularly of the tetradrachm. "In the trade of the cities of the Ægean basin and far beyond, from the beginning of the Athenian sea-power and all through it, the tetradrachms with the head of Athene and the owl took the place of native local coinages."[4] That supremacy of Athenian money vanished after the defeat of Athens. In the Western Mediterranean, the Phocæa-Mitylene combination seems to have had a similar influence during the sixth century and the earlier years of the fifth. An international currency was used in the trade of a sort of Hansa which embraced all the stations set up by the Phocæans on the coasts of Spain, Gaul, and Italy. That currency consisted of coins from the East, mostly from the monetary union of Phocæa and Mitylene, and also of coins struck by the settlers themselves on the pattern of the former.[5]

So, before the establishment of Macedonian rule, the monetary organization of Greece was marked by its variety, due to the fact that political sovereignty was divided among numbers of cities, and by the intelligent elasticity with which it was adapted to the needs of commerce. The discovery and exploitation of mines, such as those of the Laureion and Mount Pangæos, and the ever more active circulation of precious metals from Asia Minor, Colchis, Scythia, and yet more distant regions—these were conditions favourable to the issue of coinages which steadily became more abundant. Movable wealth became one of the most active factors in the economy of the Greek world; it contributed to its development

[1] *Ibid.*, p. 65. [2] *Ibid.*, p. 66. [3] *Ibid.*, p. 38.
[4] *Ibid.*, p. 60. [5] VII, 2[1], pp. 1581 *ff.*

and was itself greatly advanced by it. It went beyond the stage of the use of actual metal coins, and developed methods of exchange and payment which belong to the banking-system.

For gradually, as wealth came more and more to consist of accumulations of coin, it began to be felt that this wealth should not be allowed to lie idle, in the form of treasures jealously hoarded in temples or private fortunes locked up in cash-boxes. At the same time, the development of trade and industry drove manufacturers and merchants more and more to look for capital with which to procure materials, labour, or goods, in the hope of making a big profit from them. One man wanted an advance, and another had sufficient resources to be able to make it. The result was that the loan became a regular practice. It cannot be denied that friendly loans, without interest, were made in Greece; but it is none the less true that we find lending at interest, and at very high interest, as the usual form of this financial operation. The rate of interest might vary, according to the case, from 10 per cent. to 48 per cent. per annum.[1] The various conditions laid on the borrower and the rights and guarantees secured for the lender were specified exactly.[2]

Since, as we have seen, sea-borne trade was of the chief importance in the economic activity of the Greek world, there were many forms of contract regarding the fitting of ships and freight. One of the most characteristic was that which is usually called the loan on bottomry. "The characteristic of this kind of loan is that the lender hands capital to the borrower, to be especially applied to objects exposed to sea-risks, on condition that the borrower shall repay the sum lent only if the objects to which the loan applies arrive successfully at their destination. If, on the contrary, those objects are lost during the voyage in consequence of an accident at sea, the borrower will be released, and will not have to restore the sum lent."[3] It can easily be imagined that a loan of this kind was only granted at a high rate of interest, often for very short voyages—for example, at 12½ per cent. for the voyage from Athens to the Hellespont, and as much as 30 per cent. for a longer voyage. Minute precautions were taken to protect the lender against the

[1] Caillemer, in **XVII**, *s.v.* "Fœnus," pp. 1214 *ff.*
[2] Id., *ibid.* [3] Id., *ibid.*, p. 1220.

many frauds which the borrower might attempt.[1] In spite of the disadvantages and dangers of the bottomry loan, many Greek merchants and shipowners had recourse to it, and money-lenders did not hesitate to invest sums which were sometimes very large.

This manipulation and transmission of movable wealth, this hiring out of money (for that is what the loan at interest is, in its various forms), all these uses of metal coin are surely nothing else than the first historical manifestations of capitalism. Money was no longer merely an instrument of direct exchange between sellers and buyers; it had become a factor of economic production. It supplied the manufacturer with the means to make more goods and to make them more quickly, and the merchant with the means to offer the consumer a larger quantity of commodities, sometimes coming from more distant countries. The Athenians used to say that capital lent out at interest was active capital, working capital.[2] The expression is not only just in the sense that interest is, as it were, the product of capital; it is also just in the sense that without that capital, without its circulation and application to trade and industry, Greek economic organization would have been much more limited and much poorer.

To determine, assist, and guide these movements of capital, the direct man-to-man relations of lenders and borrowers, of creditors and debtors, were not sufficient. At an early date specialists became necessary. So banking came into existence. The first bankers, at least in Athens, seem to have been the money-changers or *trapezitai*, who originally sat in the market-place itself with their little counters or tables (*trapezai*) in front of them. Gradually they widened their activity, adding to the exchange of money a variety of financial and commercial operations, which M. Guiraud enumerates in detail as follows:

"(i.) They helped their clients to draw up contracts and took these latter into their keeping.

"(ii.) They undertook to make payments, either with sums handed over to them by the debtor or with money advanced by themselves.

"(iii.) They opened current accounts for their clients.

[1] Id., *ibid.*, p. 1221.

[2] Id., *ibid.*, p. 1215.

A young foreigner comes to Athens; he deposits the capital which he has brought with the banker Pasion, and draws it out as he needs it.

"(iv.) They issued letters of credit. An Athenian went to Miletos, and did not want to take his money with him, for fear of losing it. He paid a sum to his banker, who wrote to his correspondent at Miletos instructing him to pay a corresponding sum to the traveller.

"(v.) They lent money to individuals and, more rarely, to cities."[1]

At the end of the fifth century and the beginning of the fourth banking made remarkable strides in Athens. From the forensic speeches of Isocrates and Demosthenes, we are especially well acquainted with the house which was managed in succession by Archestratos, Pasion, and Phormion. On his death Pasion left a large fortune, estimated at about £14,000, a very great sum for the time; this shows the importance of the various operations undertaken by the Athenian banks. Some of these operations were directly connected with commerce, such as the bottomry loan. Others were of a more strictly financial kind, such as the loan at interest, the current account, and the issue of letters of credit, but even these were of great assistance to commercial transactions. Although the texts and other documents now known refer chiefly to the commercial organization of Athens, we may assume that the same or a similar organization existed in the other big cities of the Greek world. The economic life of Greece as a whole, not only of Athens, was stimulated and perfected by it.

IV

Laws and Regulations Regarding Trade

Business now took such a place in the life of the Greeks that at an early date it was made subject to laws and regulations, which no doubt varied from one city to another, but as a whole formed a real body of commercial legislation. In many cities, measures were taken to ensure honest dealing in respect of the amount, weight, and price of goods sold. Restrictions and prohibitions were enforced which seriously

[1] **XXIII**, p. 207.

interfered with freedom of trade, but were intended to protect certain interests, public or private.

Even in Athens, where freedom was the watchword, the corn-trade was very closely supervised by the State. The feeding of Attica, which was very poor in corn, was too urgent a necessity to be left entirely to private initiative and activity. "It was necessary, by internal legislation, to encourage the importation of corn and to prevent the concealment of stocks, attempts at engrossing, and artificial raising of the price. Extremely severe laws laid down the duties of the *emporoi*, or wholesale merchants, and the *sitopolai*, or dealers in large quantities, millers, bakers, and the like. No bottomry loan could be made unless it was secured on commodities of first necessity, particularly corn; no corn might be transported elsewhere than to the Piræus; not more than one-third of any cargo which had come into the Corn Harbour could be re-exported; not more than fifty 'loads' of corn could be bought at one time. The execution of these laws was entrusted to ten Inspectors of the Emporium, who had the supervision of the warehouses, and to the *Sitophylakes*, ten in number and later thirty-five, who saw that corn, flour, and bread were sold at the current price and in good weight."[1] This example, from a city in which the greatest liberty was left to the citizen, shows that even there, in the domain of public economy, the action of the individual was limited by the collective interest, at least when that interest was of great importance. No doubt abuses might ensue, and there were Greek cities which entered upon commercial operations of questionable honesty, in the nature of engrossing; but it seems that such operations were temporary, and were not so much part of a continuous policy as expedients inspired by exceptional circumstances.

The Greeks exacted customs duty, either on incoming or on outgoing goods. In Attica, the duty on imports and on exports was 2 per cent. *ad valorem*. It was a fiscal, not an economic, measure, being intended to supply revenue to the State.

As we have seen, as early as the fifth century B.C. some of the characteristics of modern commerce were taking shape in the Greek world. Marine trade acquired an amplitude

[1] XLIV, English p. 298.

hitherto unknown. Exchanges were effected not only with hard cash but by means of letters of credit. Capital began to play a definite part in economic activity. Freedom of individual action or State interference prevailed according to the city, time, and circumstances. Commercial law came into being. The great Attic orators pleaded in commercial lawsuits. The trading ports were the most active and brilliant centres of Hellenic life.

During the period extending from the beginnings of Greek history properly so called to the departure of Alexander for Asia, the economic development of the Greeks altered the conditions of life over the greater part of the Mediterranean basin, from Sicily to the west coast of Asia Minor and from the Euxine to Cyrenaica. The agricultural and pastoral land at the disposal of man was greatly increased and much better worked. Industry developed in a manner hitherto unknown: there was a constant increase in the number of raw materials, both those obtained in Greek countries and those imported from abroad, often from distant regions; the crafts were specialized and improved; and manufactured articles, both goods in common use and objects of luxury and art, acquired a value due much less to the trade cost of the raw material than to the quality and finish of the human workmanship. Trade relations became more numerous and more extensive; the invention of money gave them an easiness and elasticity which had never been afforded by the old practice of barter.

Under the beneficent influence of all this progress, the importance and results of which are hard to appreciate for a modern mind, accustomed to the powerful working of the economic forces of today, the two essential forms of all economic activity, production and consumption, developed to a remarkable extent. They acted on each other in a way which stimulated them and, as it were, made them fruitful. The producer did not work only for local needs, but for many different markets, some at a great distance. The consumer looked abroad for the satisfaction of his normal appetites and desire for comforts and superfluities. As a result, there was an intense movement of exchanges, and the land-routes and sea-routes bore a constant stream of goods, as varied as they

MAP II. GREECE AND THE ÆGEAN.

[face p. 78.

This economic development had a decisive influence on the character of property, the organization of labour, and the nature of commercial operations. Movable wealth assumed an important position by the side of landed wealth. Then what is known as capitalism made its appearance. Reserves of coin were formed, and true banks came into being. Still on a comparatively small scale, business adopted the practice of the bill of exchange—that is, of values which were in themselves fictitious, but were real in virtue of the confidence attached to them.

A series of essential stages had been passed in the evolution of economic activity as a whole. From being domestic, it had become urban, then inter-urban, and even international. The geographical area which it would henceforward cover might be increased, it might even extend one day to the boundaries of the world known to the ancients, the Οἰκουμένη; but its characteristic features were now fixed, and until the birth of machinery they would not alter in their fundamental nature.

PART II

THE ECONOMIC LIFE OF THE HELLENISTIC WORLD

CHAPTER I

THE HISTORICAL AND ECONOMIC CONSEQUENCES OF THE EXPEDITION OF ALEXANDER[1]

THE expedition of Alexander, the conquest of the East by a Greek leader, and the foundation of the Macedonian Empire, shortlived though it was—that is a collection of events of capital importance in the history of antiquity. Expedition, conquest, and foundation had, on the fate of the Mediterranean world and of all Europe, an influence which is perhaps not yet exhausted, and the study of it is more fascinating and more informative even than that of the stirring events amid which the son of Philip passed his short life.

Was the work which was then accomplished intended, or at least foreseen, by the King of Macedon? We may assert, on the strength of definite facts and characteristic evidences, that Alexander intended and was very clearly aware of the results which he obtained and those which developed after his death. A Hellenic ruler who vigorously repudiated all distinction between Greek and Barbarian, who prepared and celebrated the Susa marriages, who did away with all distinction between Macedonian veterans and Asiatic recruits in his army, who in organizing his vast empire called upon the vanquished at the same time as his own comrades in arms—that ruler, whatever the defects of his character and the excesses of his temperament may have been, gave proof of a truly creative genius. "Aristotle still said that the Barbarians were born to be slaves. He advised Alexander to be a leader to the Greeks and a master to the Barbarians; to care for the former as one does for friends and kinsmen and to

[1] **CV**, nos. 7-8; **CVII**, no. 8.

treat the latter as one does plants and beasts. That antagonism . . . had to disappear. Alexander undertook that great task. 'He ordered all,' says an ancient writer, 'to regard the world as their country . . . good men as their kin, and bad men as foreigners.'"[1]

No doubt, the great Macedonian found the ground prepared for his new policy by facts and tendencies, the true character of which had not been perceived at first. "As business men, adventurers, travellers, physicians, and, above all, mercenaries, the Greeks spread all over the world. Already more than ten thousand of them had accompanied Xenophon on an expedition against Babylon, at a time when the physician Ctesias was being heaped with honours at the court of Susa. Since then, Greek mercenaries were generally the backbone of the Persian armies. Two Rhodians, the brothers Mentor and Memnon, commanded the Persian army in its most difficult wars, thirty thousand Greeks fought for the Great King at Issos, and, until the day when he was murdered in the Caspian mountains, he had an escort of four thousand Greeks about him."[2] For all that, although circumstances may have been very favourable in some respects, the policy adopted and enforced by Alexander was none the less daring, and, indeed, revolutionary, in the eyes of the Greeks. Proof of this lies in the resistance with which it met, which he only overcame by his energetic and sometimes cruel determination.

That determination was far-seeing, and it was fruitful. The work conceived by Alexander was not damaged or retarded, in its essentials, by his premature death, nor yet by the dismemberment of his empire and the conflicts which raged all through the end of the fourth century B.C. It is true that the political result of the conquest, the creation of a single state reaching from the Ionian Sea to the Indus and from the Caucasus and Turkestan to the Sahara, was destroyed immediately after the death of the conqueror. It is true that all the plateau of Iran, with its bastions to the north and east, afterwards fell away once more from the extended Hellenic world. But the evolution, the beginning of which is marked by the decisive victory of the Greeks over the Great King, a far-reaching evolution of a profoundly human nature,

[1] **XXXI**, vol. iii, pp. 17 *ff.* [2] *Ibid.*, p. 25.

nevertheless went on over a vast theatre, which embraced Greece Proper, Asia Minor, Mesopotamia, Syria, and Egypt. In that varied and extensive domain, it affected social life and economic activity as it did religious beliefs and intellectual and artistic progress.

I

The Extension of Geographical Knowledge

It would be too much to say that before the expedition of Alexander the Greeks knew nothing of the regions lying outside the Hellenic world. Herodotos had visited Egypt up to Syene and Asia as far as the beginning of the Iranian plateau, and in the Greek colonies of the Euxine, in Babylon, and in Susa he had picked up information about the countries which lay beyond the Caucasus, Armenia, and the valleys of the Euphrates and Tigris. After Herodotos, Ctesias, who had lived at the Great King's court, had written a book on India, several fragments of which are preserved in the *Library* of Photios (*Bibliotheca*). But the Greeks could not know much about those remote regions at the beginning of the fourth century. Any ideas which they might have obtained were not only very incomplete and vague, but usually mixed up with legends and wild lies.

One result of the campaigns and residence of the army of Alexander in countries in the heart of Asia—especially in Bactriana and India—was that these indirect, incomplete notions were at once superseded by a more definite knowledge of places, men, and things. Alexander did not just go into the eastern Satrapies of the Persian Empire as he came on them; before he marched through them, he caused the chief roads to be explored, and once he was master of them he established regular communications between them and the shores of the Mediterranean. Pliny the Elder speaks of the *mensores itinerum*[1] who surveyed the roads for the King. Moreover, Alexander wanted information about the peoples with which he came into contact and those which he had to rule. He sent Onesicritos on a regular research-mission among the Gymnosophists of India, and he ordered Nearchos to recon-

[1] *N.H.*, vi, 21.

noitre the coasts of the Indian Ocean south of the Iranian plateau.

This initiative on the part of the leader inspired several of his lieutenants with a desire to tell the story, each in his own manner, of the expedition in which they had taken part. Their narratives, which were chiefly devoted to events in which they had been themselves concerned as actors or eye-witnesses, contained geographical descriptions which, if not of a methodical or general nature, were at least based on personal observation. We know that, in addition to Onesicritos and Nearchos, Callisthenes, Anaximenes, Androsthenes, Aristobulos, and Ptolemy, son of Lagos, wrote narratives of the kind. None of these works has come down to us, but fragments of them, sometimes quite long, have been copied by various geographers, such as Eratosthenes, Strabo, and Arrian. Such information, it is true, is not above criticism; Strabo realized this, and he does not hesitate to show how necessary it sometimes is to check it or even to reject it. However, it is not always useless.[1]

Moreover, this was only the beginning of a movement of genuine exploration which it is very curious to find at that time. The successors of Alexander, and the Seleucids in particular, found it useful and necessary to know something more of the outskirts of their kingdom, the regions bordering on their own domain. In the north, the Caspian, the Sea of Aral, and the surrounding country were visited by Patrocles. In the north-east, thanks to the foundation of the Kingdom of Bactriana, travellers advanced far into Central Asia, as is shown by the *Parthica* of Apollodoros of Artemita. Seleucos and Ptolemy sent embassies beyond the Indus; Megasthenes went to the court of King Sandrocottus (Chandragupta), who had become master of the Punjab after the death of Porus and Taxiles. Deïmachos penetrated to the valley of the Ganges and reached Palimbothra (now Patna).[2] One Dionysios, on instructions from the King of Egypt, travelled to the same district. In the south-east, the Arabian seaboard was visited. The Ptolemies established several ports on the Red Sea, the chief being Berenice. From these ports, daring mariners sailed forth in search of the coasts of East Africa; they went through the strait of Bab el-Mandeb, passed Cape Guardafui,

[1] xv, 1, 2-3; *cf.* i, 2, 3; ii, 1, 2 *ff.*

[2] Id., *ibid.*

explored Socotra, and perhaps went as far as Zanzibar. Later, finally, Hippalos worked out the system of the monsoons which govern sailing in the Indian Ocean, and more voyages were made to Ceylon and the Deccan.

The results of these voyages of discovery were not only of a geographical kind. The Greeks henceforward knew much more about the natural riches of these new countries, and the great routes of penetration and communication by land and sea which converged on the Mediterranean from Western and Central Asia, India, Arabia, and East Africa became familiar to them. The horizon of their economic life was pushed back to the confines of Siberia and Tibet, to the very gates of the Far East, and to the Equator.

II

The Social Consequences of Alexander's Expedition

The direct and lasting influence of Alexander's expedition was, however, confined to a more restricted area. Interesting as are the traces of Hellenism found in independent Bactriana and north-western India, the civilization known as Alexandrian or Hellenistic did not really develop outside the countries washed by the Mediterranean—Egypt, Palestine, Phœnicia, and Syria—except in the vast basin of the Euphrates and Tigris. It did not ascend the western slopes of the plateau of Iran or go beyond the mass of the Armenian mountains, which it hardly affected. But within the regions where it held its own, it reveals itself everywhere to the eye of the historian with the same fundamental character: it is a mixed civilization, born of the mingling of Greece and the East. The conquering Greek was not absorbed by Oriental life; the conquered Easterner was not enslaved, in body or in soul, by triumphant Hellenism. There was neither absorption nor enslavement, but there were mutual fusion and penetration. "Greek civilization," Droysen writes, "did not come among wild barbarians, but among peoples possessing an ancient culture of their own. It did not destroy that culture, but seized upon it with astonishment and tried to bring it into harmony with its own habits."[1] The Eastern

[1] **XXXI**, vol. iii, p. 28.

civilizations did not meet the influence of the conquerors with active or even passive opposition; they accepted it and adapted themselves to it with a good grace.

This is not the place to show how that adaptation was brought about in religion, philosophy, literature, and art. But the connexions of social and economic life are so close that we should try to see and define the evolution which then took place in the organization of human groups, in their distribution on the surface of the soil, in their relations with the nourishing earth.

In this respect, the essential contribution of the Greeks was the diffusion of city life. Alexander set the example in the course of his campaigns by founding numerous cities in every region which he conquered. Here we shall note only the most important—Alexandria in Egypt, Alexandria in Margiana, afterwards called Antioch (now Merv), Alexandria in Aria (Herat), Alexandria in Arachosia (Kandahar), Alexandria in Sogdiana (Khujand), and Alexandreschate (near Khokand in Turkestan), besides many Alexandrias in the valley of the Indus and elsewhere.

The Seleucids and Lagids founded many similar cities. The former were responsible for all the Antiochs, Seleuceias, Apameias, and Laodiceias which sprang up in Asia Minor, Syria, and Mesopotamia; the latter were the authors of the Ptolemaïses, Berenices, Philadelpheias, and Arsinoës of Egypt, Cyrenaica, and the coast of the Red Sea. Some of these new cities attained to very great political importance and prosperity—Laodiceia on the Mæander, Apameia Cibotos in southern Phrygia, close to the Pisidian border, and Seleuceia in Cilicia Tracheia; Antioch on the Orontes and Seleuceia in Pieria; Apameia in Cœle-Syria; Seleuceia on the Euphrates and Seleuceia on the Tigris. In the kingdom of the Ptolemies, Ptolemaïs on the Nile, Ptolemaïs, Berenice, and Arsinoë in Cyrenaica, and Berenice on the Red Sea enjoyed a like pre-eminence.[1]

In addition to these cities, whose very names declare their origin, other more ancient towns received colonies, either of Macedonian veterans or of Greek immigrants, which brought them a prosperity which they had not known before—Tralles in Caria, which was also called Seleuceia and Antioch, Tarsos

[1] **LVIII**, *passim*; **LXIII** (for the cities of Asia Minor); *cf.* **LXIV**.

in Cilicia, sometimes called Antioch, Hierapolis in Syria, Damascus, Ptolemaïs in Phœnicia, formerly Accho and now Acre, Edessa to the north of Syria, Nisibis at the foot of the Armenian mountains, Clysma at the top of the Gulf of Heroöpolis (near the canal from the Red Sea to the Delta, which had been restored by the Ptolemies), Coptos on the Nile, and others.

This multiplication of urban centres and their prosperity were not purely artificial. The proof lies in the effective influence which these cities exerted all round them, which struck the ancient writers. Speaking of Syria, Ammianus Marcellinus[1] says that when Seleucos I made himself master of the country he took advantage of the very dense, peaceable, and easily-ruled population to replace the country villages by towns, some of which soon became rich and powerful. In that country, then, concentration in cities took the place of dispersion over the countryside. Pliny the Elder makes the same remark about Mesopotamia, where once the only big towns had been Babylon and Nineveh; the Macedonians collected the population in cities because of the " fertility of the soil," *propter ubertatem soli.*[2] Polybios, too, mentions the great number of Greek cities which existed in Media.[3]

The places founded by Alexander and his successors were not, therefore, merely strategic points and military centres. They contributed to transforming the mode of life of the native population. According to Strabo, the inhabitants of Bactriana and Sogdiana used to lead a nomad existence before Alexander came, and their manner of life was altered by the conqueror.[4] Droysen sums up this characteristic evolution as follows: " When one sees, in countries like Mesopotamia and Syria, a luxuriant crop of cities taking the place of the instable, sometimes nomadic, Bedouin-like communities which had hitherto lived their unprogressive life there; when one sees compact agglomerations, in which the number of needs and the possibility of satisfying them develop side by side . . . one realizes what a complete change was produced by the Hellenistic foundations and how much their influence transformed the atmosphere of Oriental life."[5]

[1] xiv, 8, 5.
[2] *N.H.*, vi, 26, 117.
[3] x, frag. 27, 3.
[4] xi, 11, 10.
[5] **XXXI**, vol. iii, p. 36.

While Greece introduced city life, on a much larger scale than formerly, to the East, the Greek states founded in the East took the monarchical form of government from the old civilizations. In the valley of the Nile, the Ptolemies were the successors of the ancient Pharaohs in every respect; in Syria and on the banks of the Tigris and Euphrates, the power of the Seleucids was not essentially different from that once held by the Achæmenids. In both cases—and it was to be the same in all the Hellenistic kingdoms—the sovereign was a god; his authority was not subject to any bounds, it was not shared with any other political organ; in theory, he was the owner of his kingdom, as he was its head; all the resources and forces which it contained were at his disposal. Whether this omnipotence was exercised by the King in person or in his name by a member of the royal family, a minister, or a favourite, made no difference; the principle of absolute monarchy governed the organization and life of the Greek kingdoms of the East. "There we find the complete realization of the abstract notion of the State as being identical with the person of the King. The sole object of the State is to express that power completely and forcibly, at home and abroad. A full treasury, a militia always ready for war, an army of officials, the obedience of the subjects, the negation of any communal or corporate autonomy having any political capacity within the State, in short, the sovereign power of the King governing all society from top to bottom, unopposed, in the presence of which nothing is left to the governed but private law—that is the character of that monarchy."[1] Everything has its beginning and its end in the Court and the central government.

An organization of this kind cannot but have some influence on economic life. The Lagids and Seleucids had means of action at their disposal which Athens herself, in the days of her greatest prosperity, had not enjoyed. They could initiate schemes for public works, roads, harbours, canals, affecting immense districts. Their government was in a position to bring together, co-ordinate, and direct to a single end every local and regional activity. No doubt, both in Syria and in Egypt, many causes of weakness and decay presently began to take effect; but they told upon the political

[1] **XXXI**, vol. iii, p. 60.

life of the kingdoms rather than upon the economic life of the Hellenistic world, and the work started by Alexander and carried on by the first Successors assumed truly unprecedented dimensions under Roman rule.

III

Circumstances Favourable to Economic Progress

Other conditions, which were eminently favourable to economic progress, ensued from Alexander's expedition. Some of these conditions were of a geographical kind and origin, the others due to human action.

In the countries then annexed to the Greek world for long centuries, there were regions whose natural wealth and advantages were and have not ceased to be remarkable, and in some respects quite exceptional.

The valley of the Nile below the lowest cataracts, the vast basin of the Tigris and Euphrates from the outliers of the Armenian mountains to the common outlet of the two rivers on the Persian Gulf, Syria, and the western valleys and plains of Asia Minor offered the farmer vast areas of prodigious fertility. Compared with these new domains, what were the best districts of Greece Proper—Thessaly, Bœotia, Elis—or even Cyrenaica and Great Greece?

Over these fertile countries ran the great natural roads which connected the Eastern Mediterranean with Western and Central Asia, the Indian Ocean, and East Africa. From Armenia and the Iranian plateau there came the roads which led from Bactriana, Sogdiana, north-western India, and, beyond the territories visited or conquered by Alexander, the distant land of the Seres. Some of these roads crossed the plateaus of Asia Minor to end on the Euxine and the Ægean, and another went through Syria and reached the Mediterranean north of Phœnicia. Along the Euphrates ran the route, both by road and by river, connecting the Persian Gulf with the Mediterranean seaboard. Egypt, and especially the Delta, formed a connexion between Greek waters and the Indian Ocean.

Then there were the ports—the ancient ports of Phœnicia, among which Tyre and Sidon still held the first place, though soon to be eclipsed by Antioch and Pierian Seleuceia; Alexandria in the north of Egypt, the inspired creation of the great Macedonian; the coast-towns founded by the first Ptolemies on the Red Sea. These, for a people of sailors and merchants like the Greeks, were precious factors of economic prosperity, which could not be allowed to remain unused.

The exploitation of all these advantages was facilitated and encouraged by the policy of Alexander and his successors.

The lords of the Persian Empire had accumulated considerable masses of precious metals, either in bullion or in the form of various articles, in their palaces at Susa, Persepolis, Pasargadæ, and Ecbatana. They also possessed, thanks to the heavy taxation with which they burdened their subjects, enormous sums in specie, particularly darics. Of these two forms of wealth, the former lay idle, shut up in the Kings' treasuries; the latter chiefly served to satisfy their whims, their luxurious tastes, their sumptuous appetites, when not used to buy men's consciences in Greek lands. When Alexander came as conqueror, he seized an enormous mass of booty. In addition to what he took on the battlefields of Issos and Arbela, where the Great King's own tents had been set up, and in the course of pursuing Darius at Damascus, the historians tell us that he found 50,000 talents at Susa, 50,000 at Persepolis, 6,000 at Pasargadæ, and yet other riches at Ecbatana. We are told that he deposited 180,000 talents in this last city.[1] The intrinsic value of the talent is commonly estimated at £216; 180,000 talents therefore represent close on forty million pounds (£38,880,000 exactly).

Alexander did not allow these resources, truly fabulous for the time, to lie idle and unfruitful. He distributed large sums to the veterans of his army who were going home to Macedonia. He paid such debts as his men might have contracted in the course of the expedition. He devoted part of his treasures to restoring the irrigation channels of Chaldæa and dredging the outflow-ditches of Lake Copaïs in Bœotia. He gave 10,000 talents to many Greek cities for the restoration of temples which had fallen into ruin. He spent much on founding the cities of which I have spoken above. In brief, to use a modern expression, he put into circulation, in the form of coin, a quantity of precious metal which had

[1] Strabo, xv, 3, 6 and 9.

hitherto been immobilized and idle, he transformed unproductive treasures into working capital. "One of the most active leavens then working in that world" (the Hellenistic world) "in formation must have been the immense mass of precious metals which the conquest of Asia placed in Alexander's hands. Before the Peloponnesian War, Athens had become the great financial power because she had 9,000 talents of minted silver in her treasury on the Acropolis,[1] apart from silver and gold vessels. Now the sums in question were on a very different scale. . . . When the new royal power, which now reigned in Asia, released these buried riches, when it let them flow from its bosom as the heart causes the blood to circulate, one can easily understand that labour and trade spread them abroad, in a circulation which constantly gained speed, through the members of the Empire, which had long been ligatured and withered. One sees how by that means the economic life of the peoples, whose strength the Persian rule had sucked like a vampire, must have revived and prospered."[2]

Alexander's successors, the Kings of Egypt and Syria, took his lessons to heart. Under the first Lagids, a regular, watchful administration, productive of order and peace, brought the State considerable resources. The annual revenue, which is estimated, apart from contributions in kind, at 14,800 talents (about £3,200,000) enabled the Ptolemies to develop the prosperity of the country very quickly, by wise regulations and judiciously ordered public works. We know less about the Seleucids in this respect than about the Lagids. Their empire was less solidly united. There is, however, no doubt that Seleucos I and Antiochos I actively strove to encourage economic progress all over Western Asia, by building roads, which they endeavoured to make safe, and by founding commercial markets and ports.

Nevertheless, the intelligent use of the public resources would not by itself have been enough to extract the full value of the Eastern countries which had now come into the circle of the Hellenic and Mediterranean world. For that to happen, another leaven was necessary, the human leaven—that is, initiative and method, the signs of a free, intelligent, voluntary energy. That spirit of initiative and that sense of

[1] £1,944,000. [2] **XXXI**, vol. i, pp. 687 *ff*.

method were brought into the East by the Greeks. For a long time, travellers, historians, and philosophers had observed and pointed out the contrast which existed in this respect between Greeks and Orientals. In his true or fictitious account of the conversation of Xerxes and Demaratos at the beginning of the second Persian War, Herodotos contrasts the inert, cowardly slavery of the Great King's subjects with the Greek's passionate love of freedom and of the free play of his physical, intellectual, and moral forces.[1] Half a century after Herodotos, Hippocrates put down the indolent softness of the Orientals to their climate, and declared that the Greeks were destined by the very nature of their country to a harder-working and less routine-ridden life. Aristotle asserts that the Barbarians are born to be slaves.[2]

This contrast between Greek and Oriental may have been accentuated by the manner in which the Persians had tyrannized over the East from the sixth century to the fourth. "The great characteristic of the Persian rule which had weighed down upon the East for two centuries was that the unity of the Empire was purely mechanical. Nothing was demanded but submission. . . . The dominion was just too superficial for the governed ever to resign themselves to the loss of their independence, and there were continual revolts in many quarters, punished, it is true, by deportations and exterminations of peoples. There had never been a power less capable of ruling than the military, patriarchal monarchy of the Persians. It was based on force alone, in all its brutality. . . . Soon that kingship degenerated. The Persian people grew soft in the enjoyment of a sovereignty to which there was no counterpoise. The Satraps became like kings in their provinces, reigning as absolute despots, free of all responsibility, obeying only their own pleasure and guided only by their own fancy. New, more violent revolts of the subject nations were put down with more difficulty and cost all the more bloodshed. The situation was desperate unless help came from abroad."[3]

That help was brought by Alexander and the Greeks. Is it not remarkable that cities like Alexandria in Egypt, Antioch in Syria, Seleuceia on the Tigris, which were to enjoy

[1] vii, 101 *ff*.

[2] XXXI, vol. iii, p. 17.

[3] *Ibid*., p. 29.

such brilliant and lasting prosperity, were founded after the destruction of the Persian Empire, on sites the value of which seems to have been previously unknown? That districts in Asia Minor, far from the sea, were now, for the first time, alive with busy, wealthy cities? That it was not until the Hellenistic period that economic relations between East and West assumed the importance and vitality which made them one of the essential factors of the economic life of antiquity? It is impossible not to see here the fruitful influence of the Greek genius and of the policy of Alexander. The crowds of Greeks who settled in the East shook the old populations out of their barren apathy, and brought, in the stead of the tyrannical, selfish indifference of the Persian administration, their passion for work, their desire for progress, their thirst for adventure, and their spirit of innovation. Colonization was understood as a combined effort of the natives and the new masters of the conquered countries, the former supplying labour for every form of economic work and the latter directing, improving, and fertilizing that work, which in consequence gave an infinitely greater return.

With that collaboration, which was intended and started by Alexander and carried on by his successors, the Mediterranean East starts upon an era of activity and economic prosperity, the development, various forms, and character of which we now have to examine.

CHAPTER II

THE ECONOMIC LIFE OF THE HELLENISTIC WORLD: AGRICULTURE[1]

THE first fact to strike one on approaching the study of the period in the history of economic life which commences with the expedition of Alexander, is the increase of the actual area directly covered by that life. It is true that that area was reduced about the middle and in the second part of the third century B.C., when the eastern Satrapies broke loose from the authority of the Seleucids. While Hellenism continued to exert influence in various forms as far as Bactriana and the banks of the Indus, Iran and the adjoining regions on the north and east fell away from the Hellenistic world. Nevertheless, in spite of that loss, Hellenism extended considerably to the east and south of the Eastern Mediterranean. Henceforward it was bounded on the east by the mountains overlooking the Tigris and on the south by the Arabian desert in Asia, the strait of Bab el-Mandeb at the end of the Red Sea, and the cataract near Syene on the Nile. From now onwards, in the conquered and reorganized East, there was not a single patch of the Mediterranean seaboard which did not take part in the economic life of Greece, and further away the basin of the Euphrates and Tigris and the African coast of the Red Sea assumed a place of increasing importance in that life and played a more and more active part in it.

The result of these annexations was that the Greek peninsula and the surrounding seas fell into quite a secondary place. It has long been observed that the once thriving and glorious cities of Greece Proper—Sparta, Argos, Thebes, even Athens and Corinth—must then have made way for cities in Asia and Egypt—Pergamon, Rhodes, Antioch, Alexandria—newer, no doubt, without a great name and a past, but more alive, more populous, more brilliant. What is true of political energy and literary and artistic brilliance is no less true of

[1] For Egypt, see **LIX**, pp. 89 *ff.*; *cf.* pp. 75 *ff.*

economic prosperity. The causes of the phenomenon are many and complex. Allusion was made to some of them in the preceding chapter. It will be useful to recall them, to consider them in detail, and to supplement them.

In the fifth and fourth centuries, Greece stood, between the Ægean and Ionian Seas, at the centre of the Hellenic world. Whatever vicissitudes and political upheavals may have befallen the cities of southern Italy, Sicily, Cyrenaica, Asiatic Ionia, and the Hellespont and Bosphorus, the destinies of Hellenism were guided by the orators or Archons of Athens, Themistocles, Aristeides, Cimon, Pericles, Alcibiades, Demosthenes; by the Kings, Ephors, or generals of Sparta, Leonidas, Lysander, Agesilaos; by the Theban leaders, Pelopidas, Epaminondas; or, lastly, by the Kings of Macedon, Philip and Alexander. The old land of Greece truly continued to be the chief centre of Greek activity and power. But after the conquest of the East the situation was completely changed. All over Asia Minor, cities were founded and rose to greatness, and in them the city life characteristic of the social organization of the Greeks developed rapidly. The semi-Oriental monarchies of the Seleucids and Lagids, based on the divine essence of the King and a highly organized administration and order of precedence, presented for at least a century, in the face of the almost anarchic dispersion of energies, incessant rivalries, and disastrous jealousies of the states of Greece Proper, two organisms, more compact and more coherent, to which even the Kingdom of Macedon could not be compared.

To attract the chief economic centres to the East and establish them there, there were, in addition to these political differences, the natural riches of most of the countries which now came into the stream of Greek history. It was no longer in the plains of Thessaly and Bœotia, at the foot of the Acropolis, on the banks of the Pamisos and Alpheios, or about the seas connected by the Strait of Messina, that economic enterprise found the most fertile soil and agriculture was most active. The mining districts of the Laureion and Pangæos lost the privilege of being the principal sources of monetary circulation. The great markets deserted the shores of Greek Europe and established themselves, either on the eastern fringes of the Hellenistic kingdoms or on the Mediterranean itself, at the ends of the trade-routes of Western Asia and the Nile valley.

What still further contributed to the break-down of the balance between old Greece and the Hellenized East, was the lamentable decay into which the former fell in consequence of the discords and wars of which it was the scene right down to the Roman conquest. Diodoros and Polybios speak time and again of the looting in which the belligerents indulged—Kings of Macedon, Ætolians, Achæans, and their supporters and opponents. In the last years of the fourth century, Cassander and Polysperchon laid most of Greece waste.[1] Twenty-five years later, the invading Gauls sowed panic and ruin wherever they passed. At the end of the third century and at the beginning of the second, the Ætolians pillaged the Peloponnese, Epeiros, Thessaly, and Macedonia.[2] The Kings of Macedon ravaged Ætolia, Acarnania, Elis, Achæa, and Laconia.[3] The sanctuaries were not spared. Dodona was looted by the Ætolians,[4] the temple of Thermos in Ætolia by Philip of Macedon.[5] In addition to these particular facts, Diodoros gives us information about the general state of Greece which leaves no doubt of the wholesale ruin of the country. In 217, when peace was signed between the Ætolians on the one side and Philip, King of Macedon, and the Achæans on the other, Polybios remarks that the Achæans and other peoples of the Peloponnese could then, for the first time, return to peaceful work, repair their damages, till the ground, and hold the usual religious ceremonies and games; before that, war had been so continuous that they had almost forgotten every other form of human activity.[6] And later, speaking of the time when the Romans made Greece a province, Diodoros writes: "Since the beginning of historical times, Greece was never plunged in such calamities; it would be impossible to describe them or to read of them without tears."[7] Polybios lays stress on the shortage of men, the depopulation of the towns and desertion of the countryside, in the same period.[8]

[1] Diod. Sic., xx, 100.
[2] Polyb., iv, 3, 18, 25, 58, 62, 67; v, 17, 30, etc.
[3] Id., iv, 63 *ff*., 73, 83; v, 7-8, 18, etc.
[4] Id., iv, 67.
[5] Id., v, 8 *ff*.
[6] Id., v, 106.
[7] Diod. Sic., xxxii, frag.
[8] Polyb., xxxvii, frag 4.

The emigration of the Greeks to Eastern countries was at once a cause and a consequence of this impoverishment. Spontaneous exodus and systematic colonization organized by the Kings of Egypt and Syria contributed to draining Greece of its inhabitants. The general decay engendered intestine strife. " Plots and riots followed each other without interruption. The fewer men there were, the more fiercely they contested the land. They fought over ruins. The land remained untilled. The deserted cities fell into the condition of overgrown villages. Grass grew on the squares, and the cattle came and browsed there. Ancient Greece was at the point of death."[1]

To the East, then, henceforward vivified by the Hellenic genius, we must turn to see the development and progress of ancient economic life.

I

The Agricultural and Pastoral Resources of the Hellenized East

After the conquest of the East, a new field, not only enormous in size, but very varied in its nature and possibilities, was open to the agricultural activity of the Hellenes. From the Ægean to the plateau of Iran and from the Euxine and Caucasus to the Indian Ocean and the lowest cataracts of the Nile, there was a succession of valleys and plains of almost inexhaustible fertility, plateaus not quite so fruitful, semi-desert tracts, and wooded mountains and hills. Here broad rivers rolled, swollen by the snows of Armenia or the torrential rains of Equatorial Africa and the Ethiopian mountains; there ranges close to the sea were scored by torrents whose beds, almost always dry, sometimes filled in a few hours with a roaring mass of water. The Mediterranean climate varied; Cyrenaica was not like Egypt, nor Syria and the districts on the Ægean like the central plateau of Asia Minor. Beyond the Mediterranean frontage, Arabia Petræa and Arabia Deserta, Babylonia and Mesopotamia, Assyria and Armenia, were subject to the great variations of temperature, the excess of rain and drought (sometimes in the same region), character-

[1] XLIV, English p. 333.

istic of mountainous or continental countries and those near the Tropics.

In what economic condition did the conquering Greeks find this immense domain ? We lack the general and exact data to answer the question; we only have some scattered details, which are rather contradictory. Strabo, for example, tells us that in Media, in the time of Persian rule, horse-breeding was most prosperous, and that 50,000 mares belonging to the royal stud grazed in a meadow region called the Hippoboton.[1] Another picture is presented by Diodoros, who tells of the ruin of Sidon, which revolted against Artaxerxes Ochus and was burned down by its own defenders,[2] and of the pillaging of Egypt by the same Achæmenid King after a rebellion of that wealthy country.[3] A policy of that kind can only destroy all economic prosperity, especially agricultural. That, no doubt, is what explains the silted, blocked condition in which Alexander found the canals by which the lower basin of the Tigris and Euphrates was drained and irrigated.[4] One may, therefore, suppose that at the end of the fourth century the exploitation of the soil left much to be desired in the vast regions now annexed to the Hellenic world.

What, in this respect, were the consequences of the territorial and political achievement of Alexander and his successors ? To tell the truth, the countries which formed the kingdoms of Egypt, Syria, and Macedon and the smaller states which gradually broke off from them—Pontos, Bithynia, Pergamon, etc.—did not know the blessings of continual peace. The wars of the Successors, the conflicts of the Lagids and Seleucids, the invasion of the Gauls in Asia Minor, and revolts like that of Molon, Satrap of Media, against Antiochos III must have done harm to peaceful agriculture and pastoral life. But these wars and disturbances do not seem to have been followed by ravaging of at all a systematic kind. On the other hand, the foundation of numerous cities and the establishment of Greek settlers by Alexander and the Hellenistic kings in many regions of the old Persian Empire certainly gave a great impetus to the exploitation of natural resources, animal and vegetable. We know, indeed, that the Macedon-

[1] xi, 13, 7; *cf.* Polyb., x, frag. 4, 27.
[2] Diod. Sic., xvi, 45.
[3] Id., xvi, 51.
[4] Strabo, xvi, 1, 9-11.

ians introduced wine-growing into Babylonia and Susiana,[1] and we may presume that the remarkable development of it in Western Asia was not unconnected with that initiative on the part of the Greeks.

It is known that while Alexander was in Babylon he gave orders for the repairing of the channels, dikes, and dams by means of which the surrounding country was irrigated. In Egypt, Ptolemy II Philadelphos made the Fayum. "The most remarkable achievement of the reign of Philadelphos, or, at least, that which created most remark," writes Bouché-Leclercq, "was the creation of a new province, peopled by Græco-Macedonian or Hellenized settlers, most of them veterans, in the depression which since time immemorial had been occupied by Lake Mœris and a few Egyptian villages, the chief of which was the 'town' of Crocodilopolis. By drainage works extremely fertile land was reclaimed, where the new inhabitants introduced the cultivation of the vine and olive."[2] In the famous document known as the Rosetta Stone, one reads: "The Nile having risen very high . . . and, as it usually does, flooding the plains, the King (Ptolemy V Epiphanes) held it back in many places by reinforcing the mouths of rivers, on which works he expended no small sums."[3] The Ptolemies made use of forced labour for the building, repairing, and upkeep of the irrigation channels which were indispensable to the cultivation of a large part of Egypt.[4] This being so, one may ask whether, in other countries than Babylonia and Egypt, the existence and maintenance of waterworks of the same kind was not due either to the action of Hellenistic rulers or to the presence of settlers of Greek origin. Thus, Strabo tells us that the Chrysorrhoas, the river of Damascus, "almost entirely expends itself in irrigation ditches, having to water a very extensive district with a very rich soil."[5] Traces of a very large system of dams and other works, probably dating from antiquity, have been found all round the town of Antioch in Margiana, which was first built by Alexander under the name of Alexandria, destroyed shortly afterwards by the barbarians of the Scythian desert, and rebuilt and enlarged

[1] Strabo, xv, 3, 11.
[2] LIII, vol. i, p. 242.
[3] *Ibid.*, p. 373.
[4] *Ibid.*, vol. iii, p. 312.
[5] xvi, 2, 16.

by a number of ditches, gave the soil sufficient of the moisture which it needed.[1]

Nowhere was this characteristic more marked than at Babylon and in the low-lying plain which terminates on the Persian Gulf. Here we may well quote a long passage of Strabo, which will give us an idea of the labour done by man to fight against nature and to make her serve his needs.

> " The Euphrates overflows every year in the first days of summer, having started when the snow melts in the Armenian mountains, and rises so high that the countryside would inevitably be converted into lakes and wholly submerged, if the overflow were not led off by means of dikes and canals, as is done in Egypt with the inundation of the Nile. For that is why the canals were made. But these canals demand a great deal of labour. For the soil is so deep, so soft, and so unresisting that it is easily carried away by the current, and so the soil soon leaves the plains uncovered and fills up the ditches and blocks their mouths. In consequence, the ditches overflow in their turn into the plains near the sea, where they form lakes and swamps and reed-beds. . . .
>
> " To prevent these floods altogether is perhaps impossible, but it is the duty of a good government to do its best to remedy the evil. The remedy is this: to build weirs so as to prevent the floods from extending too far, and also to dredge the ditches and clear their mouths so as to prevent the silt from causing them to fill too much. Now, clearing the ditches is an easy matter, but building the weirs requires a great number of workers. For the soil, being very soft and unresisting, does not support superimposed masses of earth, but gives, and drags them away with it, so making it difficult to close the mouth of the ditch. The work must be done quickly, that the ditches may be closed as quickly as possible, and may not lose all their water. For if they are dry in the summer, they draw all the water off the river, and when the river is too low it cannot supply, at the right time, the irrigation which, in that burning hot country, is an absolute necessity in summer. It is just as bad if the crops are swamped by excess of water as if they are dried up for lack of it."[2]

Polybios likewise remarks on the irrigation channels which lead water from the Euphrates, so that the volume of the river is smaller in its lower course, whereas most rivers grow larger towards the mouth.[3]

In addition to these irrigation works, the water of the Euphrates fed what Strabo calls " snails," or hydraulic screws, by means of which it was led up to the terraces which held the famous hanging gardens of Babylon. These screws worked uninterruptedly under men specially entrusted with the job.[4] When Babylon was neglected by the Kings of Syria in favour of Seleuceia on the Tigris, which had been

[1] Id., xvi, 2, 16.
[2] xvi, 1, 9-10.
[3] ix, 9, fr. 43.
[4] Strabo, xvi, 1, 5.

founded by the head of the line, the new city benefited by all these works, without any change in methods being introduced.

On the whole, the natural conditions of Babylonia were remarkably like those of Egypt, where the predominant factors were the natural phenomenon of the inundation and the human works of irrigation. Just as Egypt was a long oasis extending down the Nile, so, too, Babylonia was practically an oasis, thanks to the manner in which the water of the Tigris and Euphrates was controlled. The only tree found there was the palm.

"On the whole," Strabo says, "the country is bare and the vegetation scrubby, except for the palm; the palm is very abundant in Babylonia."[1]

Further on, after extolling the richness of Babylonia in barley, he says:

"All the rest of its subsistence comes from the palm. The palm supplies it with bread, wine, vinegar, honey, and meal. It is used for making all kinds of plaited things. The smiths use the date-stones for fuel. The same stones are used for fattening oxen and sheep, after being soaked. They say that there is a Persian song, in which three hundred and sixty uses of the palm-tree are enumerated."[2]

The culture of the palm and the peculiar system of the oasis were not the only novelties which the occupation of the East introduced into the agriculture of the Greeks. According to Aristobulos and Strabo, rice was grown in Bactriana, Susiana, Babylonia, and Lower Syria.[3] Aromatic plants were cultivated in various parts of the Seleucid kingdom—amomum in Gordyene, in the upper valley of the Tigris,[4] the aromatic rush and scented reed in the waters of the Lake of Tiberias,[5] and, last but not least, the balsam-tree in the upper valley of the Jordan and in the plain of Jericho.

"The balsam-tree," Strabo writes, "is an aromatic shrub, not unlike the cytisus and the turpentine-tree. By making incisions in the bark, they draw a sap like sticky milk from it into pots, and from them it is decanted into shells, in which it coagulates. It cures headache, incipient humours over the eyes, and amblyopia in a marvellous way; for this reason it is expensive, especially since it is not grown anywhere else. . . . They also use the wood of the balsam-tree as a spice."[6]

---

1 *Ibid.* 2 Id., xvi, 1, 14. 3 Id., xv, 1, 18.
4 Id., xvi, 1, 24. 5 Id., xvi, 2, 16.
6 Id., xvi, 2, 41; *cf.* Diod. Sic., xix, 98.

Like agriculture properly so called, Oriental stock-breeding not only greatly developed certain breeds already known to the Greeks before the expedition of Alexander, but introduced them to new breeds. However prosperous horse-rearing may have been in the plains of Europe, Sicily, or southern Italy, that prosperity could not be compared to that of the royal stud at Apameia on the Orontes, where there were 30,000 mares with 300 stallions at least, and a whole world of trainers lived.[1] It was on the territory of Apameia, too, that Seleucos Nicator and his successors established their elephants, five hundred in number.[2] In the desert country between Syria and the Euphrates valley, the Arabian Scenitæ reared camels in the intervals of brigandage.[3]

So, by merely spreading over the East, the Greeks became acquainted with new sources of agricultural and pastoral wealth and accustomed themselves to methods of farming hitherto unknown.

Outside the countries which were really occupied by the Greeks and remained an integral part of the Hellenized East for long ages afterwards, other lands—Armenia, Media, Persia, the valleys of the Cyros and Araxes, the Caucasus, Albania, Hyrcania, Aria, and Margiana—some rich in corn, wine, and fruit, others in grass-land, and others in timber of every species, henceforth contributed, indirectly but effectively, to the economic evolution and progress of the ancient world.[4]

But the country which probably had the most fruitful influence in this respect was Egypt. There for thousands of years agriculture and stock-breeding had enjoyed an unexampled prosperity. Even in the Old Kingdom, paintings and sculptures, particularly the paintings in the tombs, bear witness to the great, nay capital, importance of both in the life of the country.[5] The government of the Pharaohs had maintained the fertility of the country by the intelligent control of the Nile and its periodic inundations; dams, dikes, and canals had been constructed to correct the irregularities

[1] Strabo, xvi, 2, 10. [2] *Ibid.* [3] Id., xvi, 1, 26; 3, 1.

[4] For Armenia, the plains of the Cyros and Araxes, see Strabo, xi, 14, 4; for Media, id., xi, 13, 7 (*cf.* Polyb., x, 4, 27); for Persia, Strabo, xv, 3, 21; for the Caucasus, id., xi, 2, 15; for Albania, id., xi, 4, 1 and 3; for Hyrcania, id., xi, 7, 2; for Aria and Margiana, id., xi, 10, 1-2.

[5] **LV**; **LXII**, *passim*.

of the phenomenon, to make use of its advantages, and to justify the famous phrase of Herodotos, "Egypt is the gift of the Nile." It is true that the domination of the Persians, from the sixth to the fourth century B.C., had seriously damaged the results of that work of ages. The stupidity and brutal rages of Cambyses, the frequent revolts provoked by the despotism of the Great Kings, and the violent measures of repression and pillaging which usually followed, had injured the economic organization of the country. But, bad as it was, the harm was not irreparable. To restore the old wealth, it was sufficient to ensure peace by a strong, wise administration, which took its inspiration from natural conditions, the beloved traditions and customs of the people, and the inheritance of the Pharaohs. That was exactly what the Ptolemies did, or at least the first of them, Soter, Philadelphos, and Euergetes. The building, repair, and upkeep of irrigation canals and dikes, with all the work of embankment needed for that varied and continuous task, were carefully organized and supervised by the government. Every step was taken that the soil should bear the finest harvests possible.[1] I have already spoken of the draining of Lake Mœris and the reclaiming of the Arsinoïte Nome, the present Fayum.[2] Now that Egypt had come into the stream of Hellenic and Mediterranean life, it became more and more a chief granary of the ancient world.

A great variety of crops was grown there. Cereals (wheat, barley, and durra or sorghum), flax, the vine (the culture of which was not unknown in Pharaonic Egypt, but seems to have made great progress under the Lagids, chiefly in the Delta and Fayum), oleaginous plants (in addition to flax, they were sesame, the croton or castor oil plant, and carthamus or bastard saffron),[3] vegetables (haricot beans, broad beans, chick pea, lentils, onions), fruit-trees (the most abundant of which were palms, often standing in huge groves), various aromatic plants (the produce of which was, however, less important than the perfumes imported from Ethiopia, Arabia, and India), and, lastly, two plants peculiar to Egypt,

[1] **LIII**, vol. iii, pp. 312 *ff.*

[2] Above, p. 99.

[3] The olive does not seem to have been cultivated in Egypt; olive-oil was imported to Alexandria from abroad. *Cf.* **LIII**, vol. iii, p. 264 and p. 253 n. 2.

the papyrus and the lotus—these constituted the chief agricultural wealth of the Nile valley. Some served to feed the inhabitants direct; others supplied the raw materials for various industries; others, lastly, were exported.

There was the same variety in stock-breeding. The Egyptians had herds of oxen and goats. Very fine horses were reared in the royal stables, but the real beast of burden, the fellow-worker of man in the fields and on the roads, was the ass. The poultry-yard was kept with especial care; birds of all kinds, principally ducks and geese, but also pigeons and cranes, abounded in the country districts. Bee-keeping for honey and wax was practised in the King's gardens.

What gave the rural economy of Egypt its special character was the part played in it by the Nile, similar to that of the Euphrates and Tigris in Babylonia. But in Egypt the habits of the river had been studied better, and the activity of man had been more intelligent and effective. So the valley of the Nile was the great, typical country of agricultural wealth, far more so than that of the Tigris and Euphrates.

The foodstuffs obtained from agriculture and stock-breeding were supplemented by the often considerable products of hunting and fishing. For hunting, most of our information refers to Egypt; in Asia, Strabo merely mentions some parts of Pontos, particularly Gazelonitis and the plain of Themiscyra, as being rich in game, such as roe-deer.[1] The Egyptians were very keen sportsmen. On land they hunted gazelle, wild-goat, antelope, wild-ox, hare, and sometimes hyena and lion. On the Nile and in the neighbouring fens they attacked the crocodile and the hippopotamus and caught waterfowl of all kinds—wild-duck and various wading birds—to stock their poultry-yards.[2] The great number of hunting-scenes on the monuments proves that this sport had a very big place in the life of the inhabitants of the Nile valley, and there is every reason for supposing that they were as keen on it under the Ptolemies as under the Pharaohs.

The waters of the Euxine and the straits which led to it continued, as in past centuries, to supply a great part of the fish consumed by the Greeks. The fisheries of Lake Mæotis[3] and the coast of Pontos[4] were among the most famous, and on

[1] xii, 2, 13 and 15.
[2] **LXII**, pp. 84 *ff.*; *cf.* pp. 42 *ff.*
[3] Strabo, xi, 2, 4.
[4] Id., xii, 3, 19.

the shores of the Bosphorus men carefully watched for the tunny which passed through in great shoals from the Euxine to the Ægean. In Egypt freshwater fishing was chiefly practised, with harpoon, line and scoop, lobster-pot, and seine. Of the fish caught, some were consumed fresh, while the rest were split and put to dry.[1]

The annexation of the Asiatic and Egyptian East to the Greek world, the abolition of all political, economic, moral, and intellectual barriers between the old territories of the Great King and the domain of Hellenic civilization, and the streams of reciprocal penetration which set up between the two sides, all led to a very great extension of agriculture and stock-breeding. Plants and trees previously unknown, new methods of farming imposed by climatic conditions and river-systems such as did not exist in Greece Proper, and wild and domestic animals not belonging to its special fauna—these were the characteristic elements which the Greeks learned to utilize in the countries in which Hellenism established itself and developed.

II

The Organization of Property in the Hellenistic Kingdoms

In the organization and social system of landed property the influence of the East was no less important. In Greece, as we have seen above, the essential characteristic of landed property was that it was private and individual. The restrictions placed in some cities and at various times on the full exercise of the right of ownership did not—at least, I trust that I have shown this—alter this essential character. In cities founded in historical times, whose territory we know to have been divided into lots and distributed among the citizens by the founder (*οἰκιστής*), that origin did not in practice limit the right in any way. Only two exceptions may be mentioned. At Sparta, the lots assigned to the Equals (Ὁμοῖοι) and tilled for those citizens by the Helots were regarded in theory as belonging to the State. When Athens organized the system of cleruchies in the fifth century,

[1] **LXII**, pp. 20 *ff.*

"the land distributed to the cleruchs belonged to the State, which kept the bare ownership of it, without the enjoyment, not in order to maintain the right to a share in the revenue nor to prevent hereditary transmission, but to prevent the settlers from shirking their duty of residing on their allotments by making them over or hiring them to tenants."[1] These two exceptions to the general rule are, moreover, explained by the peculiar circumstances in which the Dorians had settled in Laconia and the Athenians had created the cleruchies. Both the Dorians and the Athenian cleruchs were conquerors, occupying a conquered country, and it was their business to maintain the possession of it for the victorious city. The Spartans gave all their time and all their strength to the service of the State; the Athenian cleruchs were still citizens of Athens and had to serve in the Athenian army. These were not at all the normal conditions of land-owning in Greece, but exceptional cases, which in no way affect the private, individual character of ownership.

Secondly, the Greeks never regarded landed property as being of superhuman, divine origin. They did not hold that the earth had originally belonged to the gods and continued to do so afterwards, even from the purely theological point of view. It is worth noting that the wealth of the great Hellenic sanctuaries did not consist of vast domains so much as of movable treasures—works of art, precious metals, capital. Apollo of Delphi, Zeus of Olympia, Athene of the Parthenon, and Demeter and the Maiden of Eleusis were not great landowners by any means.

Very different was the character of landed property in the Eastern monarchies. There the sovereign "owned the goods of his subjects as he did their persons."[2] The territory over which his political sway was exercised, vast as it was, was all Royal Domain. Pharaoh and the Great King held that eminent right of ownership in virtue of their divine nature. It was because they were gods, that they were the lords of their whole kingdoms, men and soil alike. Their right of ownership was exercised in various forms, which we shall examine later. By the side of the sovereigns, the sanctuaries also owned estates, which were held by the high-priests or priestly corporations in the name of the deity, and

[1] **LIII**, vol. iii, p. 232.

[2] *Ibid.*, p. 179.

these estates were often very large; we know of several typical instances in Asia and in Egypt.

What, in respect of this economic organization, was the policy of the Hellenistic kings, Lagids, Seleucids, Attalids of Pergamon, Kings of Pontos, and Kings of Cappadocia? The creation of numerous cities, which were real Greek colonies in Eastern countries, what has been called the city-policy practised by Alexander and carried on by the Successors, led to the institution of a form of ownership which resembled the Greek system of private property without, however, losing all trace of the Oriental conception. The land granted to these new cities was regarded as being detached from the Royal Domain.[1] It was distributed among the citizens as individuals, and each of these allotments became the private property of the recipient. But their holding of it was subject to certain obligations, such as military service and the payment of one or more taxes.[2] These obligations were just what recalled the theoretical right of the State, that is, of the King. The privilege of total exemption seems to have been granted to certain cities; we know at least that this was the case with Alexandria in Egypt, where the inhabitants had the full ownership of the ground and were exempted from land-tax down to the fourth century of the Christian era.[3]

This organization of property did not become general in the Hellenistic states. Vast extents of land continued to be held by temples. Strabo mentions several cases. The sanctuary of the goddess Ma at Comana in Cataonia had a large domain, the revenues of which were taken by the high-priest.[4] At Venasa in Morimene, the temple of Zeus owned a sacred domain which brought the high-priest a yearly income of fifteen talents (over £3,200).[5] The temple of Ma at Comana in Pontos must have been as rich in land as that at Comana in Cataonia, for at certain ritual ceremonies the high-priest wore a diadem, and he was regarded as the second personage in the State, next to the King.[6] Strabo tells us that when Pompey made himself master of the country he added to the sacred domain a piece of land measuring 60 stades (nearly

[1] **XLIV**, English p. 345.
[2] **LIII**, vol. iii, p. 233; **XLIV**, *loc. cit.*
[3] **LIII**, vol. iii, p. 154 n. 1.
[4] Strabo, xii, 2, 3.
[5] Id., xii, 2, 5.
[6] Id., xii, 3, 32; *cf. ibid.*, 31, with reference to the sacred domain of the temple of Men Pharnacu at Ameria.

seven miles) in circumference, and that somewhat later another piece of land, over twelve miles round, was added to the same domain.[1] Of the sanctuary of Anaïtis at Zela in Pontos, Strabo says that the sacred domain had been greatly diminished, having been split up into a number of principalities; in the time of the Persian dominion, the high-priest had lived there in opulence and managed the sacred land as his personal property.[2] The temple of the Mother of the Gods at Pessinus brought great revenues to its priests, who had the rank of princes.[3] It is true that in this case Strabo does not indicate the source of the revenues, and does not actually say that they came from the exploitation of a sacred estate. We may, however, presume this, by comparison with the other sanctuaries of Asia Minor. To the temple of Men Arcæos on the borders of Pisidia, near the city of Antioch, a vast extent of sacred land was attached.[4]

The Egyptian sanctuaries possessed landed estates, the area of which can be estimated in some cases. The temple of Horus at Edfu owned, at the end of the reign of Ptolemy Euergetes, over 7,000 acres, and subsequently over 11,000. In Upper Egypt, the landed property of several temples, such as those at Esna and Elephantine, covered about 14,800 acres. Isis of Philæ owned the district called the Dodecaschœnos between Syene and Tachompso, representing an estate of some 600 acres.[5]

Here we have, both in Asia and in Egypt, a form of landed property which does not seem to have existed in Greece Proper.

In any case, what gives the agricultural economy of the East, and particularly of Egypt, quite a special character is the existence of immense Royal Domains, of State land. We have no details about the organization of the domains which the Seleucids had inherited from the Persian Kings. We only know that they kept huge stables, like that of Apameia on the Orontes.[6] But at least the documents give us very detailed information about the land-system of Egypt under the Ptolemies.[7]

[1] xii, 3, 34. [2] *Ibid.*, 37. [3] Id., xii, 5, 3.
[4] xii, 8, 14. [5] **LIII**, vol. iii, p. 193 n. 2.
[6] Strabo, xvi, 2, 10; *cf.* xi, 13, 7.
[7] In the kingdom of Pergamon, the greater part of the land continued to be Royal Domain, and was cultivated by the King's slaves and serfs (Rostovtzev, in **L**, pp. 375 *ff.*).

By theological rather than political law, the whole of the land of Egypt belonged to Ptolemy as it had once belonged to Pharaoh. The men who occupied it and drew their income from it, by various rights, held it from him and had to pay him various dues in return for the concession. A large part of the soil continued to be Royal Domain, the King remaining the direct owner and exercising all the rights derived from ownership over it. A difficult question arises in connexion with the estates of the temples, the Sacred Land, and the question is all the more important in view of the fact that these estates were very large and gave or might give the priesthood a wealth and power calculated to arouse the concern and jealousy of the King. Was the Sacred Land, in theory, the King's property, or was it something distinct and independent? Bouché-Leclercq decides in favour of the first view: "In virtue of his divinity and his office as priest of all the gods, the King was the lawful owner, or, if you will, the sole manager of the divine property. It belonged to the gods, of whom the King was the sole representative on earth, but not to the priests, who merely had the enjoyment of it in the measure which the King allowed."[1] The justice of this statement is borne out, in respect of the Ptolemaic period, by two definite facts.

First, in 311, Ptolemy Soter restored to the god Horus and the goddess Buto an estate which had been given to them by Pharaoh Khabbash and taken away by Xerxes. In the inscription recording this act, we find:

> "I, Ptolemy . . . I restore the territory of Patanut to the god Horus . . . and to Buto . . . as from this day, for ever, with all its villages, all its cities, all its inhabitants, all its fields, all its waters, all its four-footed beasts, all its birds, all its herds of livestock, and everything begotten or grown there, in the condition in which it was of old and with everything that was added to it by the gift of King Khabbash," and further on, "Of all this that King Khabbash gave . . . Ptolemy has renewed the gift . . . for ever."[2]

Secondly, after the draining of Lake Mœris, part of the land thus won was assigned to the temples.

Although the circumstances are not the same in the two cases, it is plain from the texts and the facts that in each case the Sacred Land was regarded as being detached from the Royal Domain. Pharaoh Khabbash and Ptolemy could

[1] LIII, vol. iii, pp. 191 *ff.* [2] *Ibid.*, vol. i, pp. 107 *ff.*

only give Horus and Buto land which belonged to them, at least in theory. So, too, since the site formerly covered by Lake Mœris had been drained as a result of the King's wish, the new land belonged in theory to the King and the estates allotted to the temples were simply concessions.

It is true that, out of respect for the gods and perhaps also to procure the good will of the priests, the Kings did not as a rule make the Sacred Land pay the tax to which other kinds of landed property were subject, which asserted the King's eminent right over land allotted at the expense of the Domain. But there were exceptions to this rule, not at all drastic ones, perhaps, but effective all the same. There was, for example, the institution of the *apomoira* by Ptolemy Philadelphos in 265 B.C. This was a tax of a sixth on certain crops, vineyards, and orchards, which had hitherto been taken by the priests. Philadelphos decreed that it should henceforward be collected by his own revenue-officers, while being still devoted to a religious cult, that of his sister Arsinoë Philadelphos, who had died recently. To reconcile the priesthood to what Bouché-Leclercq calls a kind of small financial *coup d'État*,[1] he granted it considerable subsidies. The fact remains that, in this particular case, the King's administration took the place of that of the priests. No doubt, " the exchequer pretended to respect the tradition by which the landed property of the priesthood was exempt from taxation ";[2] really there was an encroachment on the part of the King and his officials on the privileges of the temples, and that encroachment seems to prove that the King was claiming in respect of the Sacred Land rights which, if not so extensive, were at least of the same nature and origin as those which he had over the rest of the land in his kingdom.

About that matter there may be doubt and discussion; the legal position of the land granted by the State to certain persons whom the documents call cleruchs, on the other hand, is perfectly clear. These detached portions of the Royal Domain, which the holders might lease out, bequeath to their heirs, and even make over, provided that the State did not object, were liable to a number of burdens—military service and the payment of various taxes in coin or kind, some of which seem to have been permanent and normal while others

[1] **LIII**, vol. iii, pp. 193 *ff.*

[2] *Ibid.*, p. 196.

were accidental and extraordinary.[1] "In short," Bouché-Leclercq concludes, "the tenure of the soil obliged the cleruch to pay land-tax in all its forms. . . . The land thus allotted was separated from the Royal Domain in form alone."[2]

The Royal Domain properly so called, the Sacred Land, and the allotments of the cleruchs—these were the three main classes of landed property in Ptolemaic Egypt. One should add (without going back at this point to the entirely tax-free land granted to the inhabitants of Alexandria) two classes of estates, those called ἐν δωρεᾷ and those called ἐν συντάξει. The former, which were free of taxation by the special favour of the King, but may have had to pay *apomoira*, seem to have been detached either from the Royal Domain or from the Sacred Land. The revenues of the latter, according to Bouché-Leclercq's hypothesis, were intended to take the place either of subsidies granted to the temples or of salaries of officials; they were conceded, according to the case, either temporarily or in perpetuity.[3]

Various as the origin and legal position of these landed estates seem to have been, what dominates and stamps the land-system of Egypt under the Lagids is the assertion, in practice or in theory, of the King's right of eminent ownership, and Bouché-Leclercq is right in saying that in respect of landed property everything began with the Royal Domain and everything ended with it.[4] Now, in Egypt the sovereign was the personification of the State as well as the representative of godhead on earth. It was, therefore, the State which owned the soil. Save in a few cases, it was from the State that men were supposed to hold what land they possessed, and this position made them liable to burdens, the most significant of which was the land-tax in its various forms.[5] A conception of this kind was new to the Hellenic world, and introduced hitherto unknown conditions into its economic system.

In Greece itself, landed property, without changing in

[1] *Ibid.*, pp. 233 *ff.*
[2] *Ibid.*, p. 236.
[3] *Ibid.*, pp. 222 *ff.*
[4] *Ibid.*, p. 222.
[5] Were there in Egypt private properties in the full sense of the word—that is, lands whose possessors owed nothing to the State? The question is much disputed. Most scholars seem to think that there were. In any case, such estates were very rare (**LIII**, vol. iii, p. 231 n. 1).

essence, developed in a wholly disastrous manner in the Hellenistic period. "It had ceased to split up, and, on the contrary, was rapidly moving in the opposite direction. Many emigrants sold their property; families died out, and their patrimony went to swell that of the collateral branches. The 'dearth of men' inevitably led to a reconcentration of the soil, which had been broken up into minute portions. . . . The reconstitution of the big property, combined with the growing competition of foreign countries, produced disastrous results on the agrarian situation, and particularly on what remained of the small landowners. . . . The countrysides were deserted."[1] Agrarian pauperism grew worse every year.

Agricultural prosperity, which was almost dead in old Greece, developed greatly in the Hellenized Orient. The economic life of the Mediterranean world was transformed by it.

III

Methods of Farming

In Greece Proper, the methods of agriculture and stock-farming do not seem to have been very much altered, except so far as they were inevitably affected by the gradual disappearance of the small property and the desertion of the countryside. The extension of slave labour and the reduction of cornfields and vineyards in favour of pastures, if not of waste land, a return to the offensive on the part of the brush, were the usual consequences of that economic and social phenomenon.

In the Hellenized East, the working of landed property presents a more complex appearance, and in some respects a new one.

We do not know all the details of the working of the sacred domains, but we see something of it. The sanctuaries of Asia, which owned vast territories, and the Egyptian temples, to which considerable areas of Sacred Land were attached, possessed temple-slaves (*hierodouloi*) by the thousand; that at Comana in Cataonia had 6,000, that of Zeus at Venasa 3,000, that at Comana in Pontos 6,000, that of Men Arcæos near Pisidian Antioch a multitude whose exact

[1] **XLIV**, English p. 346.

number Strabo does not give. One has heard much of the female *hierodouloi*, who were sometimes dedicated to sacred prostitution; but there were also male *hierodouloi*, and one is justified in supposing that these slaves were employed in cultivating the estates attached to the sanctuaries. Two passages in Strabo seem to establish some kind of relation between the number of temple-slaves and the size of the estate.

"The Temple of Anaïtis at Zela," he says, "is today entirely under the thumb of Pythodoris. But it has suffered at the hands of many rulers, who cut down the number of *hierodouloi* and its other sources of wealth, and the territory attached to it was reduced, being broken up into a number of principalities."[1]

Further on, to contrast this diminished glory of the sanctuary with the dignity which the Persians had allowed it to enjoy, he says:

"In old days . . . the priest reigned there as absolute lord. The population in those days consisted of the multitude of *hierodouloi* and the high-priest, who lived in great plenty, surrounded by numbers of servants, and managed the sacred land as his own property."

It is very hard to believe that these *hierodouloi* did not contribute to the plenty in question by their labour. So, too, a second passage tells us that the high-priest of Men Arcæos ruled over "a multitude of *hierodouloi* and a large area of consecrated land."[2]

In addition, there were on the domains of these sanctuaries villages, and even towns, the inhabitants of which were subject to the authority of the priests. It is probable that these people, who cultivated one part or another of the domains, had to pay taxes in kind or coin to the temple administration. An Egyptian document, known as the Stele of the Satrap, which I have already quoted above,[3] gives us definite information on the point. It refers to the land in the north of the Delta, west of the Sebennytic Nome, which Ptolemy Soter is restoring to the divine pair, Horus and Buto, who had been given it by Pharaoh Khabbash and robbed of it by Xerxes.

"I, Ptolemy," we read, "I restore the territory of Patanut to the god Horus . . . and to Buto . . . as from this day, for ever, with all its villages, all its cities . . . all its waters, all its four-footed beasts,

---

[1] xii, 3, 37. [2] xii, 8, 14. [3] P. 113.

all its birds, all its herds of livestock, and everything begotten or grown there. . . . Its calves shall belong to the Great Hawks, its bulls to the face of the goddess Nebtaui, its oxen to the Living Hawks, its milk to the Glorious Nursling, its poultry to Him who is in the Sha-t and draws his own life from himself, and all that grows on its soil shall be for the altar-table of Horus himself . . . and of Buto."[1]

Here we must doubtless understand that the priests of Horus and Buto, on becoming owners of the territory once more in virtue of Ptolemy Soter's gift, shall collect big contributions in kind from the beasts reared and crops grown by the inhabitants of the cities and villages situated on it.

Religious properties in Asia and Egypt were, therefore, exploited in two ways, sometimes being managed direct, with the aid of an often very large staff of slave workers, and sometimes being leased out in some way to the free inhabitants of the towns or country districts subject to the priests.

The working of the Royal Domains appears in the most varied forms. That parts were treated as the personal property of the sovereign and worked as such is a fact not to be denied. If the Seleucids gave over the territory of Apameia on the Orontes to their war-elephants and their stables, containing at least 300 stallions and over 30,000 mares,[2] it was because they had appropriated to themselves in that part of Syria the direct and exclusive enjoyment of a very vast district. So it was, to all appearance, with the palm-garden at Babylon.[3] So, too, the Kings of Egypt had reserved for themselves a very productive palm-grove on an island in the Thebaïd, and Pliny the Elder tells us that the gardens of Jericho, which were then celebrated as being the only place on the earth where balsam grew, were likewise the personal property of the sovereign.[4]

But such cases were fairly rare. The greater part of the Royal Domains was entrusted to men called Royal Farmers (βασιλικοὶ γεωργοί), whose legal and social status is easy to determine. "The Royal Farmers were not exactly serfs. They were not attached to the soil, and their condition, which was neither hereditary nor lower than that of other Egyptians, was that of all free men. . . . Nevertheless, the Royal Land had to be tilled, and the peasants were not absolutely free to refuse their labour in the King's service."[5] Men could even

[1] **LIII**, vol. i, pp. 107 *ff.*
[2] Strabo, xvi, 2, 10.
[3] Pliny, *N.H.*, xiii, 41.
[4] **LIII**, vol. iii, p. 246.
[5] *Ibid.*, pp. 182 *ff.*

be impressed for labour on the Royal Domain in certain cases.[1] Usually, the Royal Farmers were tenants, subject not only to the payment of a rent but to certain special conditions and obligations. They could not grow what crops they pleased; two-thirds of the land had to be put under corn—wheat, barley, or durra (or sorghum). The King's bailiffs also saw that the ground was allowed to lie fallow once in three years. The rent was paid in kind, and almost entirely in wheat, the unit of measure being the *artaba*, equal to about seven gallons. The intervention and control of the royal administration were frequent. The demands of the exchequer varied according to certain circumstances, land which had rested during the previous year, for example, being taxed more heavily than other land.[2]

Equally characteristic is the monopoly of the production of oil, which the State reserved for itself. The regulations regarding it seem to have been laid down by Ptolemy II Philadelphos. The growing of oleaginous grains was not left to private enterprise. The government fixed the amount to be grown and the area to be devoted to it in each nome; it undertook to supply the growers with the required seed; it bought the whole harvest at a rate fixed by itself; it distributed that harvest among the nomes, some of which produced a surplus while others were short. This monopoly applied not only to the Royal Land but to the domains of the temples and the allotments of the cleruchs as well.[3]

So the economic and social status of the Royal Farmers was not that of serfs of the Domain, nor that of tenants who could do as they pleased provided that they paid their rent regularly, nor that of salaried workers. It had something of all these types. The lease-contract struck between the government and the farmer did not allow the land in question to be worked free of royal interference; this was exercised at every moment in various forms. Although not bound to the soil, the farmer had to submit, in almost every stage of the year's work, to the orders of the administration. We seem here to encounter a rudimentary kind of nationalization of agriculture.

The cleruchs and the free landowners, if there were any,

[1] *Ibid.*, pp. 316 *ff.*
[2] *Ibid.*, pp. 184 *ff.*
[3] *Ibid.*, pp. 253 *ff.*

were equally subject to these ties, save in respect of the oil-monopoly, and perhaps that of certain aromatic plants. The cleruch's allotment was usually too large for him to be able to farm it properly by his own personal efforts. Some of these lots were not over five acres in size, but others were as much as fifteen or twenty acres, and probably more. A papyrus found on a mummy in 1894, dating, to all appearance, from the year 240 B.C., shows us the elements of which the staff of a farm was composed. After mentioning the members of the family, the head Asclepiades, his wife Patrophila, and his four sons, Apollophanes, aged fifteen, Apollodoros, aged thirteen, Artemidoros, aged ten, and Ptolemy, aged five, the document speaks of a nurse, Cosmia, six paid agricultural labourers, Chazaros, Rhagesobaal, Ieab, Crateros, Sitalces, and Natanbaal, a shepherd, Potamon, and a cowman, Horus.[1] Potamon and Horus, who are definitely marked off from the paid labourers, may have been slaves. The most interesting thing in the text is the mention of agricultural workers specially attached to an estate. The Greek names of the members of the family justify us in supposing that Asclepiades belonged to the cleruch class.

How are we to imagine the organization of a big estate? We know that of a very large property, of the exceptional area of about 6,500 acres, in the time of Ptolemy II Philadelphos. This property, which lay in the region of what had been Lake Mœris in the Arsinoïte Nome (the Fayum), first had to undergo much alteration and improvement. Channels and dikes had to be made, without which regular irrigation was impossible; the ground had to be thoroughly cleared of undergrowth and tree-trunks had to be burned; lastly, a small town, a market, and a few sanctuaries were built to meet the material and moral needs of the farmers.

The crops, over an area of this size, were naturally very various: corn, oil-plants, vines, vegetables, flowers, fruit-trees and meadows were all to be found there. Livestock presented the same variety. Draught-animals, pack-animals, beasts intended for slaughter and for sacrifices, and wool-bearing animals were the object of sometimes minute attention. In the stalls and pastures there were oxen, horses, mules, asses, camels, sheep, goats, lambs, and pigs. Poultry were not

[1] **LIII**, vol. iii, p. 291.

neglected, especially ducks and geese. There were hives which furnished a honey with a great reputation.[1]

The farming of a big estate was done by numbers of tenants, each of whom employed farm-labourers (paid free men or slaves), herdsmen, and stable-boys. The owner's interests were guarded and represented by a whole administrative staff, an office of clerks and book-keepers. The documents introduce us to bailiffs and scribes, chief stablemen, chiefs of granaries, chiefs of herds, and foremen of gangs of workers.[2] In the time of the Pharaohs, a big estate was, according to M. Pierre Montet, a collection of dwelling-houses, stables, plough-lands, offices, workshops, and building-yards, the inhabitants of which were engaged in the most varied occupations; everything on it belonged to the owner—beasts, men, houses, and fields.[3] There is reason to think that under the Ptolemies it was the same, and that the status of men and things had hardly altered.

[1] **LXVI**, pp. 56 *ff.*, *passim*.
[2] **LXII**, pp. 126 *ff.*
[3] **LXII**, p. 382.

CHAPTER III

THE ECONOMIC LIFE OF THE HELLENISTIC WORLD: INDUSTRY

I

Raw Materials

The conquest and occupation of the East widened the field of industrial activity round the Eastern Mediterranean, as they had widened the domain of agriculture and pastoral production. Industry henceforward had at its disposal, first, the raw materials which it had employed before, but in bigger quantities and without having to go abroad for them, and, secondly, new raw materials, the use of which, doubtless inherited from the ancient civilizations of the East, introduced processes which the Greeks had not previously practised.

The corn (chiefly wheat and barley), vines, and olive-trees which grew in such abundance in parts of Asia Minor and in Syria, Babylonia, and Egypt, gave rise to a great development in the making of bread, wine, and oil, which only a famine-year could check or hamper. In Egypt, oleaginous grains were preferred to olives. The mountain ranges of Asia Minor (in Mysia, Pontos, and Cilicia), the Caucasus, and Lebanon supplemented the timber of Greece Proper and so benefited shipbuilding and joinery. The flax of Egypt, Galilee, and Colchis and the cotton picked in Upper Egypt supplied the textile industry with materials unknown to the Greeks before the fourth century. The palm in Babylonia and the papyrus in Egypt were used for a number of manufactures. Aromatic plants and trees, such as the balsam-tree of Judæa, the saffron of Corycos in Cilicia, and the storax of Syria fed the perfume-industry.

The wools of Asia Minor, Syria, Palestine, and Egypt, the purple-shells of the Phœnician coast, and the fisheries of Pontos, Syria, and Egypt were a very great addition to the similar products, alimentary, textile, or chemical, which the Greeks had formerly used.[1]

But it was above all in respect of mineral raw materials that the means at the disposal of Greek industry were increased.

In various parts of Asia the Greeks found big salt lakes and mines of rock-salt. Lake Tatta in the heart of Anatolia might be compared, in Strabo's words, to a huge natural salt-pit, because of the way in which the salt contained in its water stuck to anything which one dipped in it.[2] There were mines of rock-salt in Camisene and Culupene, in the heart of Pontos near Lesser Armenia,[3] and salt-works in the Halys valley.[4] Lake Mantiane or Matiane, on the borders of Armenia and Media, the biggest of the salt lakes after Mæotis, formed large natural deposits of salt on its shores.[5] Outside the countries which remained permanently under the sway of the Seleucids, Carmania, south of Iran, contained a mountain of salt,[6] and on the southern shore of the Persian Gulf, round Gerrha, "the ground," Strabo says, "is impregnated with salt."[7] Egypt was rich in salt, which came from lakes or marshes or from deposits of rock-salt.[8]

Nitre or natron was found in Lydia. A large lake in Armenia, named Arsene or Thopitis, produced great quantities of it; its water, which was full of it, was used for removing stains and bleaching cloth.[9] In Egypt, in the Delta, south of Alexandria and not far from Momemphis, the Nitriote Nome owed its name to two very productive nitre-beds.[10]

In Judæa, Lake Asphaltitis (the Dead Sea), which Strabo calls Lake Sirbonis, astonished the Greeks by the nature and content of its water.

"The lake is full of asphalt. At irregular periods, it spouts in the middle, from the bottom of the lake, bubbling violently like boiling water. . . . Asphalt is an earthy substance which, under the influence of heat, liquefies and springs up and expands, but, on coming into contact with cold water, like that of the lake, turns into a solid rock which

---

[1] For all these animal and vegetable raw materials, see the previous chapter.
[2] Strabo, xii, 5, 4.
[3] Id., xii, 3, 37.
[4] Id., xii, 3, 12.
[5] Id., xi, 14, 8.
[6] Id., xv, 2, 14.
[7] Id., xvi, 3, 3.
[8] **LIII**, vol. iii, pp. 238 *ff*.
[9] Strabo, xi, 14, 8.
[10] Id., xvii, 1, 23; *cf.* **LIII**, vol. iii, p. 240.

has to be cut and broken up. . . . The asphalt floats . . . and men go out on rafts and cut it and carry away as much of it as they can."[1]

Diodoros, who gives the same information about this asphalt, adds:

"The barbarians who trade in this asphalt convey it to Egypt, where it is bought for embalming the dead; for the bodies would not be preserved long if asphalt were not mixed with the spices."[2]

The income to be obtained from the Dead Sea tempted the greed of Antigonos and his son Demetrios Poliorcetes. Antigonos tried to take possession of it in 312 B.C., but the people who lived on its shores resisted and inflicted a severe defeat on him.[3]

Even petroleum, under the name of liquid asphalt or naphtha, was among the products with which the Greeks became acquainted in Assyria, Babylonia, and Susiana. Strabo speaks of naphtha-wells on the left bank of the Tigris, in the country of the Gordyæans and in the environs of Arbela.[4] They also existed in Babylonia, and, above all, in Susiana. The details given by Strabo are characteristic and worth quoting.

"If you smear a body with naphtha and put it near a fire, it bursts nto flame. It cannot be extinguished with water (which only makes i t burn the more) unless a very great deal is poured on. The flame can only be put out with mud, vinegar, alum, or birdlime. They say that Alexander, as an experiment, poured naphtha over a slave in the bath and put a torch to him. The slave at once blazed up and came near to perishing, only the bystanders put out the fire with torrents of water and so saved his life. Poseidonios states that in Babylonia there are wells of white naphtha and others of black naphtha, and that the former contain liquid sulphur (these are the ones which attract fire), while the latter yield liquid asphalt, which is burned in lamps instead of olive-oil."[5]

The iron-mines of the country of the Chalybes south of Trapezus,[6] those of the district which Strabo calls the Realgar Works (Sandaracurgeion), south of Sinope, which consisted of long galleries cut in the sides of a mountain;[7] the copper-mines of Cyprus, famous for their richness;[8] the deposits of red-lead or Sinope earth in Cappadocia[9] and Carmania;[10] the

1 xvi, 2, 42.
2 Diod. Sic., xix, 99.
3 Id., xix, 100.
4 Id., xvi, 1, 4 and 24.
5 Id., xvi, 1, 15.
6 Id., xii, 3, 19.
7 Id., xii, 3, 40.
8 Id., xiv, 6, 5.
9 Id., xii, 2, 10.
10 Id., xv, 2, 14.

gold of Colchis and Armenia;[1] the silver of Carmania;[2] the tin of Drangiana[3]—all these ores and metals were added to those extracted and used in Greece Proper before the time of Alexander. New stone-quarries in Cappadocia[4] and Egypt[5] and the marbles of Mylasa and Synnada[6] furnished precious materials to the architects and sculptors of the Hellenistic age.

II

The Chief Centres of Manufacture

The industries properly so called, that is, the various manufactures fed by all these raw materials, developed and multiplied just as the extraction and gathering of the raw materials had increased and progressed. Some of these industries had been practised before, but now rose to heights previously unknown; others came for the first time into the Greek economic sphere.

The manufacture of food is everywhere indispensable to human life. It does not, on the whole, seem to have centred especially in any cities and assumed particular importance there. The ancient authors do not tell us of any flour-milling centre. Sinope, on the coast of the Euxine, was renowned for the manufacture of olive-oil.[7] Babylon made sesame-oil.[8] In Egypt, all oleaginous grains—sesame, castor-oil-plant, carthamus, flax, etc.—were treated, and supplied oil for food, lighting, perfumery, and embalming.[9] Barley, which was harvested in large quantities in the Nile valley, served for the manufacture of a drink something like beer; since this was to some extent the national drink of the Egyptians, there were many busy breweries in the country.[10]

At an early date the Greeks had practised the industries by which fish is preserved—drying, salting, smoking. In many countries of the Hellenized East these trades developed greatly. On the Euxine, in addition to the old Greek colonies on Lake Mæotis, most of the cities on the north coast of Asia

[1] Id., xi, 2, 19; 14, 9.
[2] Id., xv, 2, 14.
[3] Id., xv, 2, 10.
[4] Id., xii, 2, 9-10.
[5] **XXXV**, pp. 17 *ff*.
[6] Strabo, xii, 8, 14; xiv, 2, 23.
[7] **XXXV**, p. 44.
[8] Strabo, xvi, 1, 14.
[9] **LIII**, vol. iii, pp. 253 *ff*.
[10] *Ibid.*, pp. 248 *ff*.

Minor, from Chalcedon on the Bosphorus to Dioscurias in Colchis, owed at least part of their prosperity to them.[1] Nowhere were they more active than in Egypt, where, however, only freshwater fish was used, sea-fishing having always been neglected in that country. However, the salt fish of Egypt supplied an export trade.[2]

The textile industries reached a very high degree of prosperity in the Hellenistic period. In Greece itself they had very much declined, though wool was still worked in Achæa, chiefly at Patræ, and the introduction of flax-growing into Elis had led to the creation of cloth-mills in that district.[3] But all over the East linen, wool, and even cotton and silk were spun and woven. The island of Amorgos in the Ægean continued to manufacture linen celebrated for its fineness.[4] Linen was woven in Colchis.[5] The reputation of the linens of Cilicia, although chiefly attested for the first centuries of the Christian era,[6] may have been older. Laodiceia in Syria, Byblos, Berytos, and Tyre in Phœnicia, and Scythopolis in the Jordan valley were centres of the linen industry.[7] Borsippa, near Babylon, was the centre of a great manufacture of linen-cloth.[8] The same industry was one of the most ancient in Egypt, where almost all the people wore nothing but linen.[9]

The weaving of wool and the manufacture of woollens employed hosts of workers all over Asia Minor. No longer was Miletos the only place to work in wool, as in the past. Lydia and Sardis, Phrygia with its famous embroideries, Galatia, and Cilicia, which manufactured a felt woven of goat-hair, exported the products of their textile industries.[10] Cyprus, most of the Phœnician towns, and Damascus manufactured not only common woollens, but carpets of various colours and tapestries adorned with fantastic animals.[11] This was also the special characteristic of the woollen industry in Egypt, chiefly in Alexandria; it produced far less ordinary

1 **XXXV**, p. 42.
2 *Ibid.*, p. 14; *cf.* Diod. Sic., i, 36 and 52.
3 **XXXIV**, *Prov.*, English vol. i, p. 292.
4 See above, p. 54.
5 **XXXV**, p. 43.
6 *Ibid.*, p. 30.
7 *Ibid.*, pp. 19, 24, 26; **XXXIV**, *Prov.*, English vol. ii, p. 137.
8 Strabo, xvi, 1, 7.
9 **XXXV**, p. 6.
10 *Ibid.*, pp. 27 *ff.*, *passim.*
11 *Ibid.*, pp. 19, 26, 53.

stuffs than fancy materials, adorned with mythical scenes and other subjects.[1]

Was cotton grown in Egypt and Phœnicia? It is difficult to say so definitely. Cotton-weaving, at least, existed in both countries.[2]

The island of Cos was celebrated for its silk tissues, and it is even probable that silk-worms were bred there.[3] Phœnicia, too, manufactured silk goods, but it is supposed that it imported either the cocoons or the thread from Central Asia and the Far East.[4]

Lastly, woollens and silks were woven or embroidered with gold at Sardis, Pergamon, and Alexandria.[5]

Closely connected with the textile industries went the production of purple, dyeing, and the manufacture of colours. Before the conquest of Asia by Alexander, murex-fishing and the production of purple were practised in many cities on the west coast of Asia Minor and in several of the neighbouring islands—Cos, for instance. Cyprus and Phœnicia were particularly important centres of the industry. From Phœnicia, where it originated, it spread in Palestine to Sarephtha, Dora, and, later, Cæsareia, and to Neapolis and Lydda in the interior.[6] Egypt possessed dye-works and manufactured colours.[7] Strabo gives us a piece of information about the very plentiful waters of Hierapolis, near the Mæander, which complements the above data:

> "For dyeing wool, the water of Hierapolis has marvellous virtues; woollens dyed in that city with simple roots vie with those dyed with kermes and purple."[8]

From this passage one may suppose that Hierapolis was an important dyeing centre. Furthermore, the water of Lake Arsene or Thopitis, which was full of nitre, was excellent for removing stains and bleaching cloth.[9] There may have been bleaching-works on the shores of the lake.

While in Attica and Eubœa the mines of the Laureion and

1 *Ibid.*, p. 10.
2 *Ibid.*, pp. 10 (Egypt), 19 (Phœnicia).
3 *Ibid.*, pp. 48 *ff.*
4 **XXXV**, p. 19.
5 *Ibid.*, pp. 36 (Sardis), 38 (Pergamon); **XXXIV**, *Prov.*, English vol. ii, p. 255 (Alexandria).
6 *Ibid.*, English p. 137.
7 **LIII**, vol. iii, p. 270; **XXXV**, p. 14.
8 Strabo, xiii, 4, 14.
9 Id., xi, 14, 8.

Chalcis were almost abandoned,[1] at Corinth[2] and Delos[3] bronze-working was still prosperous, and bronze-casting continued to be active at Chios and Samos.[4] Cyzicos and Rhodes manufactured weapons, and Pergamon became an important metallurgical centre.[5] In the extreme south of Phrygia, near the frontiers of Lycia and Pamphylia, the people of Cibyra were renowned for their skill in working and chasing iron.[6] The iron-industry was especially developed east of Pontos, in the country of the Chalybes, which contained the most plentiful and best worked mines in all the East; the name of *Sinopikon* given to the steel which Sinope exported proves that iron-ore was worked in the district.[7] Although not so prosperous as in Homeric times, the Phœnician goldsmiths and metal-chasers continued to supply vases of gold, silver, and bronze, jewels, tripods, and candlesticks to many distant and barbarous peoples, as far as Britain in the west and the shores of the Indian Ocean in the south.[8]

Shipbuilding had fallen off sadly in Greece Proper, even at the Piræus, for Athens had to send a deputation to Antigonos in 307 to beg him for materials for the purpose;[9] but it developed in certain Asiatic ports—at Cyzicos,[10] at Sinope,[11] in Cilicia,[12] in Cyprus,[13] and in Phœnicia, chiefly at Tyre[14]—that is, wherever there were large forests within reach.

Many industries which in modern statistics appear under the head of chemical industries were known to the Greeks in Hellenistic times.

The potters no longer created, as they had done in previous centuries, the painted vases which were works of art rather than industrial products. They turned out, by the thousand, the elegant, delicate statuettes, some of them realistic to the point of caricature, the best-known of which have been found at Tanagra in Bœotia, Myrina in Æolis, and Tarsos in Cilicia.[15] Above all, the prosperity of the vineyards, the reputation of certain brands, and the development of the wine-trade led to the manufacture of increasing quantities of common

1 **XXXV**, pp. 68, 87.
2 *Ibid.*, p. 74.
3 *Ibid.*, pp. 91 *ff.*
4 *Ibid.*, pp. 45 *ff.*
5 **XXXV**, pp. 39 (Cyzicos), 50 (Rhodes), 38 *ff.* (Pergamon).
6 Strabo, xiii, 4, 17.
7 **XXXV**, pp. 40 *ff.*
8 *Ibid.*, pp. 19 *ff.*
9 Diod. Sic., xx, 46.
10 **XXXV**, p. 39.
11 *Ibid.*, p. 43.
12 *Ibid.*, p. 31.
13 *Ibid.*, p. 52.
14 *Ibid.*, p. 20.
15 **XLIX**, pp. 79 *ff.*

vessels, particularly amphoras, for transporting the precious fluid. Cassandreia in Macedonia, Chios, Samos, Rhodes, and Cnidos were renowned for their potteries.[1]

Pottery had long been familiar to the Greeks; it was in the East that they were initiated into glass-making. Phœnicia and Egypt were the chief centres. It is difficult, not to say impossible, to say in which of the two countries the industry was older, and whether it was borrowed by one from the other or was invented and developed independently in both.[2] However that may be, in both countries the chief raw material of glass-making was to be had on the spot.

"Between Accho (Ptolemaïs) and Tyre," Strabo tells us, "the coast is a sandy beach containing *hyalitis* (vitrifiable sand). On the actual spot, they say, the sand does not fuse, but when it is taken to Sidon it vitrifies easily. Some, however, say that the Sidonians also have this *hyalitis* sand, which is suited to fusion, and yet others declare that all sand, everywhere, fuses. When I was in Alexandria, I was told by the glass-workers that there is a vitrifiable earth in Egypt too, and that without it they could not make their valuable many-coloured glassware."[3]

Sidon in Phœnicia and Alexandria in Egypt especially went in for this industry. Egyptian glassware was probably more artistic than Phœnician.[4]

The manufacture of perfumes and unguents was, like glass-making, of Oriental origin. If we except Cyzicos and Cyrene, where we hear of perfume-works,[5] all the centres of the industry were in the East—at Tarsos, Soli, and Corycos in Cilicia,[6] and in Cyprus,[7] Phœnicia,[8] Syria,[9] Palestine,[10] Babylonia,[11] and Egypt.[12] The raw materials were sometimes obtained on the spot, flowers at Cyrene, saffron in Cilicia, storax and lilies round Antioch and Laodiceia in Syria, and balsam at Jericho. Often, too, they were imported from

[1] **XXXV**, pp. 57 (Cassandreia), 45 (Chios), 46 *ff.* (Samos), 51 (Rhodes), 33 *ff.* (Cnidos and environs).

[2] *Ibid.*, p. 11.

[3] xvi, 2, 25.

[4] **XXXV**, pp. 11 *ff.*

[5] *Ibid.*, pp. 39 (Cyzicos), 5 (Cyrene).

[6] *Ibid.*, p. 30; the Cilician perfumers chiefly used saffron, which was grown in the plain of Corycos.

[7] *Ibid.*, p. 54.

[8] *Ibid.*, p. 20.

[9] *Ibid.*, p. 26; the chief Syrian perfumes were those extracted from the plant called *styrax* (storax) and the lily.

[10] *Ibid.*, p. 25; the balsam of Palestine was especially renowned.

[11] *Ibid.*, p. 26.

[12] *Ibid.*, pp. 13 *ff.*; **LIII**, vol. iii, pp. 242 *ff.*

foreign and sometimes very remote lands, Ethiopia, Arabia, India.

Even more local than the manufacture of perfumes and glass were two industries whose importance cannot be exaggerated—that of parchment at Pergamon and that of papyrus in Egypt. Pliny relates that one of the Ptolemies, out of jealousy of the library at Pergamon and to encourage that of Alexandria, forbade the exportation of papyrus. Then it was that the King of Pergamon, Eumenes II, developed and popularized a process already known, which consisted in preparing the skins of animals for writing. So began the industry of parchment (Περγαμήνον, *membrana*). But parchment was much too costly, and never spread as papyrus did. It seems to have been reserved for rare and valuable *volumina*.[1] With papyrus the Egyptians made many other things than paper—small boats, sails, clothing, carpets, sandals, riddles, and ropes. But all these articles only served for their own use, whereas paper soon acquired great economic value and became a very important article of exportation. As in glass-making, so in paper-making Alexandria stood first of all the cities of Egypt. It was in the Delta that the papyrus-plant was chiefly to be found.[2]

At Babylon and in Babylonia the palm-tree played a part comparable to that of the papyrus in Egypt.[3]

This long list of various industries shows how important industrial labour became after the expedition of Alexander. In Greece itself this form of human activity seems to have slackened, but most of the Eastern countries enjoyed an undeniable prosperity in the Hellenistic age. The textile industries, metal-working, shipbuilding, pottery, and the manufacture of glass, perfumes, parchment, and paper kept or acquired a very important place in Asia Minor, Phœnicia, Syria, and Egypt. In many cities the raw materials required were sometimes imported from great distances, and the manufactured goods, which far exceeded the needs of local and regional consumption, fed an extensive trade.

Two documents, one relating to Syria and the other to Egypt, give us an idea of the development of certain industries, of their technical perfection, and of the costly nature of their

[1] Pliny, *N.H.*, xiii, 70; **XXXV**, pp. 87 *ff*.
[2] *Ibid.*, pp. 12 *ff*.
[3] Strabo, xvi, 1, 14.

products. These are the accounts of the ceremonies known as the Pomp of Antiochos Epiphanes at Daphne and the Pomp of Ptolemy II at Alexandria.

Of the Pomp of Antiochos, Polybios has left us a detailed and accurate picture. Antiochos Epiphanes wanted to outdo the magnificent games which Æmilius Paullus, the vanquisher of Perseus, had held in Macedonia, and accordingly sent round word on all sides that he was going to hold great games at Daphne, near Antioch. The prelude to these games was a Pomp (πόμπη) which was as follows. At the head went thousands of soldiers, recruited in Macedonia and Thrace or in Asia Minor; foot-soldiers armed with bronze or silver shields and wearing gold crowns. They were followed by squadrons of cavalry, clad in purple cloaks, some of which were covered with plates of gold, while others were embroidered with figures of living creatures. After the cavalry came chariots, each drawn by six or four horses, and others drawn by four or two elephants. Then came thirty-six elephants, in full equipment. Behind this military pageant went a hundred youths with gold crowns, a thousand splendid bulls, nearly three hundred deputations sent by cities, eight hundred elephant-tusks, countless statues of gods, goddesses, and heroes, and an incalculable mass of gold and silver plate. One friend of the King, named Dionysios, was represented in the procession by a thousand young slaves, bearing silver vases, none weighing less than a thousand drachmas. The King himself sent six hundred, who carried gold vases. Eighty women, magnificently dressed, rode in litters with gilded legs, and five hundred in litters with silver legs. During the games, which went on for thirty days, a variety of perfumes was scattered about from golden vases—saffron, cinnamon, nard, fenugreek, amaracus, and iris. Banquets were served, at which the diners sat sometimes on a thousand couches and sometimes on fifteen hundred, all sumptuously decorated.[1]

The Pomp of Ptolemy II Philadelphos, the most likely date for which is 279 B.C.,[2] was described by Callixenos of Rhodes in a work which was preserved in the library of Alexandria but is now lost; the text of his account has been preserved by Athenæos.[3] Its magnificence has perhaps been

[1] xxxi, 1, fr. 3. [2] **LIII**, vol. i, pp. 155 *ff.*
[3] v. 25 *ff.*, pp. 196A-203B; *cf.* **LIII**, vol. i, p. 155 n. 2.

exaggerated; but it bears witness to the industrial and artistic wealth of the kingdom of the Ptolemies. "There was in the procession a huge car laden with silver plate, among which one saw a mixing-bowl holding 600 *metretai,* very artistically worked and covered with precious stones, two sideboards, ten large basins, sixteen mixing-bowls, a table twelve cubits in size, thirty others of six cubits, eighty Delphic tripods, and an infinite number of other articles, all of solid silver. Then came the car of gold utensils, which included twenty-two coolers, four big gold tripods, an altar three cubits high, and, above all, a gold casket, adorned with precious stones, ten cubits high, containing six compartments, which were decorated with numbers of figures of beautiful workmanship, four palms high. By the two vehicles walked 1,600 children, 250 of whom carried gold *congii,* 400 silver *congii,* and the rest gold and silver coolers, etc. On another car there were a gold thyrsus, 90 cubits long, and a silver spear of 60 cubits; another bore a gold phallus 120 cubits long, not to mention quantities of gold utensils, vases, and weapons (including 64 complete suits of armour); and there were twenty other cars laden with gold, four hundred with silver, and eight hundred with spices. In the King's tent, at the place where the meal was laid out, there was gold and silver plate worth 10,000 talents."[1] To these almost fabulous riches we must add the costumes of the children and others, the equipment of a body of over 100,000 foot-soldiers and 20,000 horsemen, and a parade of rare or exotic animals, elephants drawing vehicles or mounted, camels, ostriches, etc. The sceptical Bouché-Leclercq says that this fantastic procession puts a severe strain on the credulity of the most indulgent reader.[2] It is quite possible that Callixenos of Rhodes exaggerated the number and value of the objects which were displayed on this occasion. He lived some sixty years later, and so could not be present; he does not speak as an eye-witness.[3] All the same, his picture shows what an impression was produced by the magnificence of the show.

In these two feasts at Antioch and Alexandria, which were separated in time by a hundred years, what strikes one is not the intrinsic value of the precious metals lavished on both

[1] **XXXI,** vol. iii, p. 52. [2] **LIII,** vol. i, p. 156.
[3] *Ibid.,* p. 155 n. 2.

occasions; it is also, and chiefly, the fact that the things exhibited for the crowd to admire were products of human workmanship. Arms, harness, vases of every kind and shape, litters, couches, clothing, tissues, and perfumes owed their value not only to the materials of which they were made, but also, and chiefly, to the skill and talent of the artists and craftsmen who had modelled, chased, woven, embroidered, forged, and fashioned them.

This prosperity contrasts vividly with the lamentable poverty into which Greece Proper had sunk at the same time. Polybios several times lays weight on the ruin of the Greek cities. He shows us the King of Macedon, Philip, in 219, lacking corn to feed his troops and money to pay them, and begging both of the Achæan League, which gives him fifty talents, as pay for three months, and 10,000 bushels of wheat.[1]

III

The Organization of Labour

The development and prosperity of the industries in the Hellenized East resulted in, or at least were accompanied by, remarkable progress in the division of labour and the specialization of workers. Certain trades were handed down from father to son. "At every turn," M. Glotz writes, "we meet obscure lines of farmers, masons, carpenters, blacksmiths."[2] For centuries it had been so in Egypt; according to Diodoros, regulations were issued by the Pharaohs making every profession a hereditary duty. Here the ancient historian may perhaps have transformed what was only an inveterate habit into an obligation, but the fact remains that in practice every Egyptian who plied a trade had learned it from his father and taught it to his son in his turn; in this way professional traditions were preserved.[3]

The result was that the worker became more and more shut up in his own occupation, and this organization of industrial work in watertight compartments grew more marked as time went on. "The Hellenistic period carried specialization very far. At Delos the joiner who fits a door does not set up the

[1] Polyb., v, 1, 6 *ff.*
[2] **XLIV**, English p. 321.
[3] **LIII**, vol. iii, p. 167.

post which is to hold it. Before the carpenter lays upon the top course of a wall the elm plank which is to support the cross-beams of the ceiling, the mason is called in to level that course. The stone-masons attached to the temple do not sharpen their own tools. At Miletos the workmen who do the fine cutting of the marble squares of the facing are not the men who rough-hew the stone blocks of the inner core. It has been possible to draw up from the papyri an endless list of the trades practised in Egypt. . . . The sack-carrier and the milk-carrier are quite different from the baggage-porter. . . . The forges and pot-banks have their oven-men and stokers. The quarryman, the man who cuts the stone, refuses to sweep away the sand or to remove a layer of gravel; it is not his job. . . . In the oil-mills there are grain-crushers, and special workmen for castor oil; garments of common cloth and fine *othonia* are neither made nor sold by the same people; and the textile trade includes a wool-waste-gatherer. Everyone to his trade."[1]

With this specialization went improvement in technical methods. The future workman was no longer trained in the family alone. The apprentice system was organized, as is proved by the papyri.[2]

For all these different industries, labour was supplied by various elements of society. In the mines and quarries, the work, which was very hard, was done by convicts.

> "The Realgar Works (*Sandaracurgeion*)," Strabo writes, "are a mountain" (south of Sinope, west of the Halys) "which has been hollowed out by mining operations, since the miners went into it by deep cuttings. The lessees who used to work it obtained their men among the slaves sold because of crimes."[3]

In Egypt, the miners who worked for the King were regular chain-gangs of convicts, guarded by soldiers.[4]

The use of slaves as workmen was still the rule.[5] The slave-trade was now at its height. The Cilician pirates were the chief purveyors.

> "It was easy for them to procure prisoners of war, and not far away there was a big and wealthy market, that of Delos, which could take in and send out tens of thousands of slaves in a day."[6]

---

[1] **XLIV**, English p. 321.
[2] *Ibid.*, English pp. 351-52.
[3] xii, 3, 40.
[4] **LIII**, vol. iii, p. 241.
[5] See above, pp. 58-9.
[6] Strabo, xiv, 5, 2.

No doubt most of the slaves sold at Delos were taken to Rome, Italy, and the West; but it is likely that a good number of these unfortunates remained in the East and that the workshops of Egypt and Asia Minor absorbed at least some of them.[1]

Craftsmen of free birth and condition were numerous in the big towns of the Hellenized East. The important thing to note here is the fact that they combined in trade-guilds. These guilds seem to have been very widespread in Egypt.[2] At Thebes we hear of three guilds of the Openers, Embalmers, and Guardians of the dead; at Alexandria, the durra-crushers and the stevedores of the port formed two self-governing corporations; and similar organizations of dockmen perhaps existed at Berytos and Pierian Laodiceia.[3] There is reason for believing that this tendency to form professional associations was general in the kingdoms of the Lagids, Seleucids, and Attalids, where industry was greatly developed and large urban centres attracted an ever-increasing population of manual labourers.

The methods of managing a business varied as much as the forms of labour. That there were individuals who ran concerns of their own everywhere is obvious; it is a universal phenomenon. What is more peculiar is the part played by the temples and the sovereigns.

At Didyma, sacred slaves were employed on the extraction of marble and the building of the temple of Apollo.[4] In Egypt, the sanctuaries often had industrial establishments attached, particularly oil-mills and weaving-works.[5] "The priests had to make the garments of the gods with their own hands. . . . The temples had to make the finest tissues, the *byssi* which were in such demand all through Greek and Roman antiquity."[6] In these workshops sacred slaves and free men worked together. Did the great sanctuaries of Asia Minor, those at Zela, the two Comanas, and Pessinus, possess a similar organization? We do not know.

More important than the workshops of the priesthood were those of the Kings. We know those of Egypt and the

[1] **LXI**, p. 106. [2] *Ibid.*, p. 105.
[3] **LIII**, vol. iii, pp. 167 *ff.*, 172.
[4] **XLIV**, English p. 351; *cf.* **LVII**, pp. 158 *ff.*
[5] **LIII**, vol. iii, p. 205. [6] **LXI**, p. 78.

Kingdom of Pergamon best. In these two states, the King kept for himself certain industrial monopolies, both in the extraction of natural products and in manufacture. In Egypt, the working of quarries and mines, nitre-beds, and salt-deposits, and the manufacture of oil and textiles were royal monopolies,[1] and so, perhaps, was the paper-industry.[2] There was a similar state of things among the Seleucids and Attalids. The Seleucids exploited "the wealth contained in the soil," by which we must doubtless understand quarries, mines, deposits of rock-salt, etc. "The Kings of Pergamon . . . possessed tile-works and brickfields, their parchment competed with the Ptolemies' papyrus, and they placed their beautiful gold-plated fabrics on the market."[3] This organization is even found in some Greek cities. At Miletos there were municipal weaving and dressmaking works, and elsewhere, at Rhodes, Cnidos, and Smyrna, pottery was a State industry.[4]

These various monopolies were not everywhere conducted in the same manner. In one place there were real factories, managed by officials, in which the work was done either by salaried free workers, as seems to have been the case with the textile and oil industries in Egypt,[5] or by slaves, as at Pergamon and Miletos.[6] In another place the monopoly was farmed out to contractors, controlled by government officials; quarries and mines were worked in this way, and probably also salt-pits and nitre-beds.[7] In addition, for certain public works of general interest and sometimes of an urgent kind, such as the making and repair of canals and dikes in Egypt, the royal administration had recourse to the *corvée*, or forced labour.[8]

The State also interfered in industry by enforcing detailed regulations. We know in particular those governing the manufacture of oil and paper in Egypt. At every stage of the industry, from the area in each nome to be sown with the chief oil-plants to the manner in which the oil was to be

[1] **LXI,** pp. 62 *ff.*, 88 *ff.*; **LIII,** vol. iii, pp. 238 *ff.*, 253 *ff.*
[2] *Ibid.*, p. 267.
[3] **XLIV,** English p. 355; *cf.* p. 353.
[4] *Ibid.*, same pp.
[5] **LXI,** pp. 70 *ff.*; **LIII,** vol. iii, pp. 259 *ff.*
[6] **XLIV,** English p. 351.
[7] **LIII,** vol. iii, pp. 289 *ff.*; **LXI,** p. 90 (salt).
[8] **LXI,** p. 128; **LIII,** vol. iii, p. 312.

sold, we are confronted with minute regulations.[1] Paper could only be made in sizes which were strictly specified.[2]

Lastly, a characteristic sign of the conception prevailing in the matter of industry, at least in Ptolemaic Egypt, is the existence of a State tax on the trades, the *cheironaxion*. " Craftsmen " (and merchants) " paid individually a special tax for the right to practise their trade. . . . The *cheironaxion* was a tax on the trade itself. It had to be paid for the licence to ply that trade. . . . All craftsmen were liable to this tax. Even the humblest trades did not escape; we hear " (in the documents) " of baggage-porters, donkeymen, and masons. . . . For each trade there was a special rate. . . . Second-hand dealers, menders, felt-makers, and wool-carders gave their names to so many special taxes; so did carpet-makers and cobblers. . . . The tax was quite independent of the earnings, custom, and scale of business of the particular merchant " (or craftsman); " in each class, and there must have been as many classes as trades, everybody paid the same amount."[3] The principle, the underlying reason of this tax, was doubtless analogous to that of the land-tax; like every tiller of the soil, every handicraftsman was supposed to hold the right to ply his trade and earn his living from the sovereign, who was regarded as the ultimate master of industrial production as of the land. The connexion seems to have been seen by Strabo, when he lumps together in one sentence " all who in peace-time till the soil and practise a trade, and supply the King with revenues by their taxes."[4] If this is indeed the meaning of the tax on trades, it points to a control of economic matters by the State of an extreme kind, which appears to be derived from the essential character of the Oriental monarchy and to have had no place in the former civilization of Greece.

The development of industry in the Hellenized East does not seem to have bettered the material condition of the worker. To say nothing of the condition of the convicts in the quarries and the slaves, public or private, the life of the free worker was very hard, especially in Greece Proper. There the labour problem was extremely serious. Wages fell and unemployment was frequent. Disputes constantly arose

[1] *Ibid.*, pp. 254 *ff.*
[2] *Ibid.*, p. 267.
[3] **LXI**, pp. 105 *ff.*
[4] xvii, 1, 3.

between workers and employers. The former went on strike; the latter tried, on pretexts of varying speciousness, to evade the payment of the wages which they had promised. "So the working classes had much to endure, but far more in the districts of Greece Proper, which had no raw materials, than in the Eastern countries, where the building of great cities and the riches of the soil favoured industry. . . . We can understand why so many workers left their country and exchanged their tools for arms, at the prospect of the fine pay offered by the kings. The brilliance of Hellenistic civilization covers more misery than can be reckoned."[1]

[1] **XLIV**, English pp. 356 *ff.*, 361.

CHAPTER IV

TRADE IN THE HELLENIZED EAST

THE commerce of Greece and the East underwent as great a change after the conquest of Alexander and the diffusion of Hellenism in the valley of the Nile and all Western Asia as agriculture and industry. In fact, the progress of these latter forms of human activity contributed greatly to the change.

Although the regions which were divided among the kingdoms of the Successors never came to form the single huge state which Alexander had projected and even realized, they kept up daily, constant relations with one another. Over all the national or local civilizations inherited from a long and often brilliant past, Hellenism was a factor of unity in every domain. One can, therefore, treat the whole Hellenistic world as one economic whole, complex, no doubt, but having all its parts mutually connected—all the countries comprised by the Adriatic and the Ionian Sea in the west, the Sahara, Ethiopia, and Arabia in the south, Iran in the east, and the Caspian, the Caucasus, the plains of Scythia, and the wild mountain-country of the Balkans in the north.

From one end of this area to the other, the land was not all equally fertile or barren, raw materials were not equally plentiful, and human activity was not equally intense, skilful, or successful. There were valleys and plains of amazing fertility, and there were districts which were quite or almost quite uncultivated. By the Nile and the Euphrates there were endless cornfields, which were lacking in Greece Proper, the islands, and many parts of Asia Minor. The olive, which was found almost everywhere in Greece and in many parts of Asia, did not grow in Egypt. The tablelands and valleys of Asia Minor supported countless flocks of sheep, which supplied wool which was much in demand. Timber, which was plentiful on forest-clad mountains like Ida in Mysia, the Taurus in Cilicia, the Caucasus, and Lebanon, was wholly lacking along the Nile and in Babylonia. The iron-mines of the country of the Chalybes and the copper-deposits of Cyprus were the only sources of these metals in the East. The big industrial centres were distributed at intervals, from Sinope and Cyzicos in the north of Asia Minor to the Phœnician cities and Alexandria. The great economic phenomenon of division of labour and specialization, not only between individual human beings but between natural regions, developed to a hitherto unknown extent in the now much wider domain of Hellenistic civilization. The consequences were inevitable: for every foodstuff, for every raw material, for every manufactured article, commercial currents set up between the centres of production, or rather of over-production, and the countries which consumed without producing or consumed more than they produced. It is true that Greece had known movements of exchange of this kind before Alexander's time, but under the Seleucids and Lagids these movements became far more extensive. Every kind of goods now travelled across the Hellenized East and old Greece.

This movement was not confined within the limits by which the political dominion of the Greeks was ultimately bounded. The work of Alexander had broken down the barriers which had previously separated the Mediterranean world from the inner parts of Asia. Those barriers were not set up again when Iran, Bactriana, Sogdiana, and the upper valley of the Indus broke off from the empire of the Seleucids. Not only did Greeks and Hellenized Orientals go to all these countries for many foodstuffs and raw materials required by their industries, not only did they export their manufactured goods to them, but they were able to use the great routes which ran through those regions, connecting the shores of the Mediterranean with Central Asia and the Far East in one direction and with Arabia in another. At the same time, the ships of Ptolemaic Egypt began to throng the Red Sea, to pass the strait of Bab el-Mandeb, and to navigate the Indian Ocean, either along the coasts of southern Arabia and East Africa or direct to the Deccan. In this way a very extensive and truly international trade grew up, and for the first time the economic life of the ancient world came under the influence of Southern and Eastern Asia and of parts of the seaboard of Equatorial Africa.

The actual organization of the Hellenistic kingdoms dominated this commercial activity as it dominated agricultural and industrial life. Great as was the influence once exercised by the markets of Athens and the Piræus on the development of business, it cannot be compared, even remotely, to that of the Kings of Syria and Egypt. By the foundation of numerous cities at well-chosen spots, by an urban policy inspired by the example of Alexander, by their wise interest in the public works needed for the building and upkeep of ways of communication of all kinds, by their interference (perhaps too frequent and a little tyrannical, but useful) in the circulation of goods and exchanges between individuals, the Seleucids, the Attalids, and above all the Lagids showed themselves capable of taking advantage of the very favourable new conditions offered to commerce. Their subjects profited by those conditions more fully than private enterprise could have done by itself, with its necessarily limited means of action.

All these causes acted in the same direction. In spite of the wars and disturbances too often provoked by the rival ambitions of kings and the jealousies of cities, in spite of the decline into which the royal houses of Alexandria, Antioch, and Pergamon fell after a hundred years, the eastern basin of the Mediterranean now lived a very intense economic life, involving regions which had long been almost unknown to one another and now found their interests closely allied.

I

The Great Trade-Routes by Land, River, and Sea[1]

Of the roads which ran all over the countries of Hellenistic civilization, the most important were those connecting the shores of the Mediterranean with the regions lying away from its basin.

The northernmost of these roads was that which crossed the Caucasian isthmus, connecting the basin of the Cyros and Araxes with that of the Phasis and coming out at the eastern end of the Euxine.[2] From Sarapana, " a fort which

[1] **CVI**, pp. 12-13.

[2] Strabo, xi, 2, 17.

can hold the population of a city," travellers and goods went down the Phasis by water.

Through the middle of Asia Minor and the kingdom of the Seleucids ran the great route by which Sogdiana, Bactriana, and the upper valley of the Indus communicated with Mesopotamia, Syria, and the seaboard of the Ægean. It was in a way the backbone of the road-system which covered Western Asia. Strabo describes big stretches of it in various parts of his *Geography*. From Ephesos to Samosata on the Euphrates it ran, on the whole, apart from some detours, from west to east, by the valley of the Mæander, southern Phrygia, Lycaonia, Cappadocia, and Commagene. On this stretch it touched Magnesia, Tralles, Nysa, Antioch on the Mæander, Laodiceia on the Mæander, Apameia Cibotos, Burnt Laodiceia (*Katakekaumene*), Coropassos, and Mazaca.[1] From Samosata, it doubtless went down the Euphrates to Babylon and Seleuceia on the Tigris; from there it traversed Media to the pass known as the Caspian Gates.[2] Beyond the pass, it first kept on the northern slope of the plateau of Iran as far as Arian Alexandria (Herat), and then divided. One branch went straight on through Bactriana to Ortospana among the Paropamisadæ, in the upper valley of the Cophen, a tributary of the Indus; the other diverged southwards through Drangiana, passed through Prophthasia, and then turned east and went onwards to the Indus.[3] From Ortospana a road started which ran to Bactra and Sogdiana; this was the road by which the caravans came from Central Asia and the country of the Seres.[4]

North and south of this vital artery, other roads served more local traffic. We must suppose that there was a route which ran from the valleys of the Tigris and Euphrates through western Media and Armenia to the Caucasus, which it crossed either by the Pass of Darial, the Caucasian or Sarmatian Gates of the ancients, or by the Albanian Gates near Darband on the coast of the Caspian, and went on to the country of the Aorsi, not far from where the Don and Volga approach one another; for Strabo says that the Aorsi had the monopoly of the camel-transport of goods consigned from India and Babylonia by way of Armenia and Media.[5] Droysen thinks

[1] Strabo, xiv, 2, 29. [2] Id., xi, 13, 7. [3] Id., xv, 2, 8.
[4] **XLIV**, English p. 375. [5] xi, 5, 8.

that this lucrative traffic, which was chiefly in the hands of the Armenians south of the Caucasus as it was wholly in those of the Aorsi north of that range, must have had its terminus at Seleuceia on the Tigris.[1]

Moreover, Seleuceia and Babylon, situated on the Tigris and Euphrates respectively in the district where the two great rivers came close together, constituted a very important road-centre. They stood at the junction of roads from upper Persia through Susiana and from Arabia Felix through Gerrha and Chaldæa.[2] There was also active intercourse with Syria. The itinerary followed by merchants travelling from Syria to Seleuceia and Babylon is thus described by Strabo:

> "They cross the Euphrates at the level of Anthemusia, a place in Mesopotamia. Above the river, four *schoinoi* away, stands the city of Bambyce, which is also called Edessa and Hierapolis. . . . Then, after crossing the river, they go over the desert to the Babylonian frontier, and so come to Scenæ, a big town on a canal. The distance from the Euphrates crossing to Scenæ is reckoned at twenty-five days' journey. There are camel-men who keep khans, which are well supplied with water, mostly from cisterns but in some cases brought to the place. . . . Scenæ is eighteen stades from Seleuceia."[3]

The road from the Euphrates to Syria by Palmyra does not seem to have been used in Hellenistic times. The importance of Palmyra dates from the Roman Empire.[4]

From Arabia and Palestine, there were roads to Syria and Phœnicia. Caravans from Arabia Felix went to Damascus, but for a long time the route which they took was infested by brigands.[5] Eastern Palestine and Transjordania were connected with the port of Arados by a road which followed the valleys of the Jordan and Lycos.[6] South-east of Palestine, Petra was an important road-centre. It stood on roads coming from Arabia Felix and Leuce Come, the port of the Nabatæans on the Red Sea, and on one coming from Gerrha across the desert, and it was connected with Rhinocolura, a town on the Egyptian frontier between Gaza and Pelusion. Strabo speaks of Rhinocolura as a Phœnician city, and adds that the goods conveyed from Petra to it were sent on from

[1] **XXXI**, vol. iii, p. 72.
[2] *Ibid.*; for the road from Gerrha, *cf.* Strabo, xvi, 3, 3; for the transport of spices from the country of the Sabæans to Mesopotamia, id., xvi, 4, 19.
[3] xvi, 1, 27.
[4] **XXXIV**, *Prov.*, English vol. ii, pp. 92-93.
[5] Strabo, xvi, 2, 20.
[6] Id., xvi, 2, 16.

there in every direction.[1] Yet other roads served for the transport of spices from Arabia Felix to Gerrha on the Persian Gulf or to Ælana at the head of the Bay of Ælana.[2]

In Egypt, the land-routes chiefly served to maintain communications between the valley of the Nile and the coast of the Red Sea. When Ptolemy II Philadelphos had founded several stations on that coast, the most important being Berenice, almost exactly at the latitude of Syene, it was necessary to connect them with the river on whose banks the life of the country centred. Coptos, at the point where the Nile came nearest to the Red Sea, stood at the head of a number of roads from Myos Hormos, Leucos Limen, and Berenice. The biggest and most used was that from Berenice.

> "There is a kind of isthmus between Coptos and the Red Sea at Berenice. . . . Philadelphos, they say, first built this road, using his troops for the work, and established stations along it. . . . Experience has proved its usefulness, and today all goods from India and Arabia and all Ethiopian goods which come by the Red Sea are taken to Coptos. . . . In the past merchants used to travel on their camels by night . . . taking their water with them; today there are watering-places along the road, both wells dug to a great depth and rain-water cisterns, although rain is rare in that country. This route is six or seven days long."[3]

In old Greece, no attempt seems to have been made to extend or improve the road-system. In Macedonia a few stretches of road were built, but it was not till later that the Via Egnatia ran continuously from the Adriatic to the Ægean.[4]

The conveyance of goods by land was not an unknown thing to the Greeks before Alexander's time, although they far preferred the sea. On the other hand, they had practically never attempted transport by river, except perhaps to reach their trading-station of Naucratis in the Nile Delta. The Hellenized East possessed waterways of great importance, which had been used for ages and long continued to be used under the Kings who succeeded Alexander.

From the passage of Strabo, quoted above, regarding the road which crossed the Caucasian isthmus south of the mountains, it would seem that the Phasis was navigable up to Sarapana, which is usually placed in the upper valley of the

[1] Id., xvi, 4, 24.
[2] Id., xvi, 4, 4. For overland connexions between Arabia Felix and the regions near the Eastern Mediterranean, *cf.* **XXXIV**, *Prov.*, English vol. ii, p. 151 and n.
[3] Strabo, xvii, 1, 45.
[4] **XLIV**, English p. 370.

river. The Cyros, which flowed into the Caspian, may also have been navigable for some distance, since the same author states that the carriage-road which started from Sarapana led in four days to the banks of the Cyros. May one not suppose, from the words used by the geographer, that this overland road was simply a portage over the mountains between the waterways of the Phasis and the Cyros ?[1]

On the Euphrates and Tigris, navigation was very active, at least on their lower waters. Vessels (by which one must understand those which had sailed up the Persian Gulf to the adjoining and almost common mouths of the two rivers) could go on the former to Babylon and on the latter to Seleuceia and even Opis, a big trading centre about seventy miles upstream from Seleuceia as the crow flies.[2] In addition, rafts, which the people of Gerrha used for the transport of spices and other foodstuffs coming from Arabia, went up the Euphrates to Thapsacos, and it was not till there that their cargoes were unloaded, to be sent overland to their various destinations.[3]

When Alexander arrived in Chaldæa, shipping on the lower Euphrates and Tigris had been obstructed by the Persians for strategical reasons.

> " From fear of invasion from outside, to make it difficult to ascend the two rivers, they had made artificial cataracts. When Alexander came, he destroyed as many of these as possible, and especially those below Opis."[4]

The conqueror was also at pains to ensure the normal working of the many canals which had been dug all over Babylonia. These canals were chiefly used for irrigation, but shipping also took advantage of them, for Strabo says that it is equally injured by extreme drought and by too high water, and the only remedy in either case is to be able " to open or close the mouths of the canals quickly, so as to keep the water always at an average level and not to allow there to be either too much or too little."[5] There was a canal connecting the Euphrates with the Tigris, which it joined at Seleuceia; but whether it was made by the Seleucids or existed before Alexander's time is not known. In any case, it seems certain

[1] Strabo, xi, 2, 17.
[2] Id., xvi, 1, 9.
[3] Id., xvi, 3, 3.
[4] Id., xvi, 1, 9.
[5] xvi, 1, 10.

that the Greeks did not remain content to use the water-system, natural or artificial, which existed before they settled in the country; Alexander and perhaps the first Seleucids improved it and doubtless also completed it.

Similar work was done by the Ptolemies in Egypt. Here the Nile was the great road of transport from Syene to the sea, and, as in Babylonia, the canals, which were chiefly intended for irrigation, were used by cargo-boats. The upkeep of these canals and their sluices and the embankments by which they were contained like the Nile itself was ensured by the forced labour of those who dwelt on their banks.[1] The Lagid government never neglected these works, which were indispensable to the economic welfare of the kingdom. The construction of the roads from various ports on the Red Sea to Coptos gave a great impetus to Nile-borne traffic between that city and the Delta. In the Delta itself shipping reached an intensity never known before. After mentioning the two outermost arms of the Nile, the Pelusiac on the east and the Canopic on the west, Strabo goes on:

> "Between these two mouths there are many others, five of them large and the rest smaller; for the first two arms give off a great number of branches, which divide up the whole Delta, forming many streams and islands; and the whole of it is navigable, since connecting canals have been cut, which are so easily navigated that on some of them there are ferry-boats made of earthenware."[2]

This development of river transport in the Delta was one consequence of the foundation of Alexandria, which soon attracted all goods to itself and was the point at which imports from abroad started for the interior.

To complete the system of waterways formed by the Nile and the canals disposed along it, the Ptolemies accomplished an undertaking which had been conceived by the Pharaohs and resumed by Darius, but for various reasons had never been finished—the construction of the canal connecting the Nile with the northern end of the Gulf of Heroöpolis, at the top of the Red Sea between Egypt and Sinai.

> "Another canal," Strabo writes, "opens into the Indian Ocean, or rather the Red Sea, near the city of Arsinoë or Cleopatris, as it is sometimes called. It passes through the Bitter Lakes, which were once bitter, but when the canal was cut were changed by the admixture of

[1] **LIII**, vol. iii, pp. 312 *ff*.

[2] Strabo, xvii, 1, 4.

water from the Nile, and are now full of fish and waterfowl. The canal was first cut by Sesostris, before the Trojan War, or, some say, by the son of Psammetichus, who was only able to begin it before he died, and subsequently by Darius I, resuming the task. He abandoned it when on the point of finishing it, under an erroneous impression; for he was persuaded that the level of the Indian Ocean was higher than Egypt, and that if the whole isthmus between them[1] was cut through Egypt would be submerged by the sea. However, the Ptolemies cut right through, and made a closed passage, so that one could sail out to the sea or sail in again, when one wished, without difficulty."[2]

To these general statements Diodoros adds a detail, namely that it was Ptolemy II Philadèlphos who took on the completion of the work commenced by Necho and Darius. An Egyptian document discovered at Heroöpolis, known as the Pithom Stele, tells us that it was finished between 280 and 264 B.C.[3]

This canal joined the Pelusiac Arm some distance below the point of the Delta, north of Heliopolis. By it, and the other canals which connected all the other branches, and the Canopos Canal, Alexandria could communicate with the Red Sea by water all the way. The Mediterranean was connected with the Indian Ocean, the Inner Sea with the Outer Sea, as they were called, not, as today, across the Isthmus of Suez, but by the Nile and by the Delta.

The importance assumed by land-routes and inland navigation in Hellenistic trade in no way diminished the marine activity which was so dear to the Greeks. Not only did their ships continue to range the Eastern Mediterranean; they ventured forth on the Indian Ocean. In the Eastern Mediterranean, regular lines of shipping grew up between Alexandria, the Phœnician ports, Rhodes, Delos, Corinth, Byzantion, and the Euxine. Westwards, the currents of trade ran chiefly towards Taras, Syracuse, Carthage, and certain points in central Italy and southern Gaul. On the Indian Ocean, the two chief directions followed by traffic, to Arabia Felix and to the Deccan, started from the east coast of Egypt and the mouths of the Tigris and Euphrates respectively. The former first ran due south, to the strait of Deire (the present

[1] Not that of Pelusion, now the Isthmus of Suez, but the stretch of land between the Gulf of Heroöpolis and the Pelusiac Arm of the Nile, in which the canal ended.

[2] Strabo, xvii, 1, 25; *cf.* Diod. Sic., i, 33. This closed passage was simply a lock.

[3] **LIII**, vol. i, pp. 159, 241.

Bab el-Mandeb), and then, turning east, followed the southern coast of Arabia and that of Gedrosia to the mouths of the Indus.[1] The latter went down the Persian Gulf from north-west to south-east and rejoined the former after passing the coast of Carmania. Lastly, Egyptian vessels were beginning to frequent the coasts of East Africa, visiting the Cinnamon Country, turning Spice Cape (Guardafui), and perhaps going some distance towards the Equator.

In the time of the Lagids and Seleucids, Mediterranean trade suffered from the scourge of piracy. The Euxine was infested with corsairs, Heniochi, Zygi, and Achæans, who came out from the rocky creeks dominated by the Caucasus,[2] and Georgians of the Tauric Chersonese.[3] A king of the Cimmerian Bosphorus named Eumelos made constant war on them for some years at the end of the fourth century. Diodoros says that he cleared the Euxine of these pirates and succeeded in protecting shipping;[4] but Eumelos only reigned five years, and it is likely that piracy again broke out in those waters after his death. In the south of Asia Minor, the coast of Cilicia became, like that of the Caucasus, a regular lair of pirates, and not till Pompey and the Romans came was an end put to their activities.[5] Yet other pirates, Cretans, Epeirots, and Illyrians, ranged the Mediterranean.

In spite of these dangers, the sea-routes played a very big part in Hellenistic commerce, and many ports on the Eastern Mediterranean attained a prosperity which they had not known before.

II

The Great Trading Centres. Entrepôts and Seaports

In old Greece, the chief centres of trade shifted eastwards, like economic life as a whole. For Athens and the Piræus, the days of active life and greatness were gone; they were approaching their death. Corinth alone, with its two ports, Lechæon on the Corinthian Gulf and Cenchreæ on the Saronic

[1] The system of the monsoons was not known before the time of Nero, and it was not till then that mariners dared to sail across the open sea from the Bab el-Mandeb to the Deccan.

[2] Strabo, xi, 2, 12.

[3] Id., vii, 4, 6.

[4] Diod. Sic., xx, 25.

[5] Strabo, xiv, 5, 2.

Gulf, and its position on the narrow isthmus between the Ionian and Ægean Seas, kept its old importance as an emporium or entrepôt of goods.[1] The famous *diolcos* over which ships were dragged from one sea to the other was still in use.[2] The capture and burning of the city by the Romans in 146 B.C. destroyed its prosperity, or rather interrupted it for a century. Corinth was revived by Julius Cæsar and Augustus, and under the Roman Empire it once more became a very busy trading port and entrepôt.

Delos was the chief gainer by the temporary eclipse of Corinth. Even before the disaster of 146, the holy island of Apollo was a port of call for most vessels proceeding from Italy or Greece to Asia.

> "The religious gathering is something like a market, and people, especially Romans, were in the habit of going there even when Corinth was standing."[3]

On Delos lines of shipping from Egypt, Macedonia, and the West converged. Phœnicians, Egyptians, and Italians had counting-houses there and had founded societies which were at once religious brotherhoods and commercial clubs. The marine activity of Delos received an immense stimulus when the island was handed over by Rome to the Athenians, on the condition that they established a free port there. At that time the great port of the Ægean was Rhodes, which exacted duty on all goods entering or leaving the harbour. The creation of the free port of Delos gave the island commercial supremacy, at least for some time. Besides, "Delos offered every security to shipping and every convenience to trade. The anchorage was protected against the north wind by a strong breakwater. A mole divided it into two parts. On one side was the Sacred Harbour, intended for the caïques which carried the pilgrims; the landing-stage with which it was provided adjoined a big meeting-place of streets and an agora. On the other side was the Merchant Harbour, where the heavy cargo-boats came in. It was divided into portions, bounded by stones, and was fringed with wharves on to which warehouses and stores opened. . . . All around stretched the commercial quarter, with shops, workshops, bazaars,

[1] Id., viii, 6, 20. [2] Polyb., iv, 19, 7 *ff.*
[3] Strabo, x, 5, 4.

and hostelries all heaped together. . . . Delos, now become 'the common emporium of the Greeks,' collected all the products of Oriental Greece, from Egypt to the Euxine, in order to forward them to Italy. . . . She organized the uninterrupted passage of slaves, grain, spices, etc."[1]

On the west coast of Asia Minor, the first place, formerly held by Miletos, was now taken by Ephesos, which Strabo describes as "a general entrepôt of the goods of Italy and Greece."[2] The port of this city, however, had serious disadvantages, largely due to ill-conceived works undertaken by Attalos Philadelphos, King of Pergamon (159–138). The mud of the Cayster had silted up the old harbour. In 287 Lysimachos had dug a new one, more to the west. Attalos Philadephos

"had imagined that the entrance to the harbour and the harbour itself would be deep enough for big vessels . . . if a mole were run across part of the entrance. . . . Exactly the opposite happened. Being kept inside the mole, the silt filled the harbour with shoals to the very mouth, whereas in the past the ebb and flow of the tide had to some extent carried the deposit away to sea. That is what the harbour is like; but, thanks to the other advantages of its situation, the city grows larger every day, and is the biggest commercial centre in Asia west of the Taurus."[3]

Ephesos stood at the end of the great road which ran over all Western Asia from Bactriana, India, and Iran.[4] From the south another road came in, from Physcos, a port in the Rhodian Peræa, by Lagina, Alabanda, and Tralles.[5]

North of Ephesos, Smyrna, which was revived by Antigonos and Lysimachos, was a busy centre; so was Miletos, to the south, notwithstanding its decline; and still further south Cnidos had two harbours and a shipyard containing slips for twenty vessels.[6]

But during the Hellenistic period the commercial queen of the Ægean was for many years the city of Rhodes. Founded in 408 at the north-eastern end of the island, the city of Rhodes had soon eclipsed the three ancient cities of Lindos, Ialysos, and Camiros. Her prosperity had in no way suffered by the rule of the Queen of Caria, Artemisia II, nor by the events through which she had passed at the time of

1 **XLIV**, English pp. 338-39; **LVI**, pp. 31 *ff.*; **LXVII**, pp. 10 *ff.*
2 Strabo, xii, 8, 15.
3 Id., xvi, 1, 24.
4 See above, p. 142.
5 Strabo, xiv, 2, 29.
6 Id., xiv, 2, 15.

Alexander's campaign. "The state of Rhodes, favoured as it was by a most fortunate geographical position, had become extremely flourishing even in Alexander's lifetime, and still more so during the wars of the Successors. Almost all the trade between Europe and Asia concentrated on the island. The Rhodians were distinguished seamen, with a reputation for honesty and skill. Their strong, constant, law-abiding character, their knowledge of business, and their admirable marine and commercial laws made Rhodes a model among all the trading cities of the Mediterranean. By her continual and successful wars with the pirates, who at that time often disturbed the peace of the seas in great bands, Rhodes had become the protectress and refuge of merchant shipping in Eastern waters."[1] Her chief trade was with Egypt, whence she received not only corn but also goods from Arabia and India, which she reconsigned to Greece and the West. It is possible that as early as the end of the fourth century she had entered into relations with Rome.[2] At the time of the wars of the Successors, the aim of her policy had been to keep neutral; it was to defend this policy that she opposed Antigonos and his son Demetrios Poliorcetes in 305.[3] The marine and commercial activity of Rhodes did not slacken for a century and a half; it only began to decline after 166, when the Athenians, at the instigation of Rome, made Delos a free port. All through the third century and the first half of the second, she ruled the Mediterranean, not only by the power of her fleet, by the extension of her trade as far west as the coasts of Gaul, Spain, and North Africa, and by her public and private opulence, but by her laws and regulations regarding marine matters.

> "Her laws are admirable, and so are her institutions in general, especially in marine matters. . . . At Rhodes, as at Marseilles and Cyzicos, everything connected with the shipbuilding yards, war-engine factories, arsenals, etc., is treated with especial attention, and even more so than in the two other cities."[4]

It was chiefly to transit commerce that Rhodes owed her great wealth. All goods entering or leaving the port had to pay customs duty. It has been reckoned that the State

[1] **XXXI**, vol. ii, pp. 448 *ff.* [2] Polyb., xxx, 5, 6.
[3] Diod. Sic., xx, 81 *ff.*; **XXXI**, vol. ii, pp. 450 *ff.*
[4] Strabo, xiv, 2, 5.

great ports of the Levant today. The movement of business was remarkably intense.

The position which Alexander had chosen for the city made it possible to build a sea-harbour on the north and a river-harbour on the south. The sea-harbour, between the shore and the island of Pharos, protected from the eastern swell by Cape Lochias and the breakwater which ran out from it, was divided into two by the Heptastadion, a mole connecting Pharos with the land. On the east of the mole was the Great Harbour, the mouth of which was marked by day and lit by night by the famous Pharos tower which was the ancestor of the modern lighthouse. On the west was the Eunostos, which was more especially a merchant harbour, whereas the Great Harbour contained the royal dockyards and war-fleet. Two openings in the Heptastadion allowed vessels to pass from one harbour to the other. The Great Harbour was so deep that the biggest ships could come alongside the quay.

Important as the sea-harbour was, the river-port played quite as big a part in the business of the city. It stood on Lake Mareotis, and communicated by the many canals of the Delta with the Nile and the Red Sea. Now, Strabo tells us that more goods came to Alexandria by these canals than by sea, and the river-port soon became richer than the sea-port. The two were connected by a canal. The chief element in the trade of Alexandria was transit; it was truly the city where Mediterranean and Indian Ocean, Europe and Asia, West and East met.

No other city in Ptolemaic Egypt could compare with Alexandria for commercial activity. It would, however, be going too far to ignore the places where the products of Arabia, India, and East Africa arrived—Berenice and Myos Hormos on the Red Sea, and, above all, Coptos on the Nile.

> "All goods from India and Arabia," Strabo writes, "and all Ethiopian goods which come by the Red Sea are taken to Coptos, and that city is the general entrepôt of such merchandise."[1]

Coptos was connected by road with the Red Sea ports.[2]

So, from the northern shore of the Euxine to the lower valley of the Nile and from the Ægean to the south-western slopes of Iran, sea and river ports, entrepôts and markets,

[1] Strabo, xvii, 1, 45. [2] See above, p. 144.

were the busiest centres of national and international commerce. All these ports, entrepôts, and markets had dealings with one another, and the goods which they received or sent out sometimes came from distant countries, of which the Greeks had scarcely heard. The commercial activity of Hellenistic times covered a very vast domain, much vaster than that of which the Greeks had obtained the mastery in the great days of their expansion and colonial power.

III

Forms and Articles of Hellenistic Trade. Trade Regulations. Modes of Exchange

In these trading centres, these entrepôts and markets, and on the great routes by sea and land which covered the Eastern Mediterranean and all the neighbouring countries, quantities of goods of every kind were brought together, exchanged, and sent abroad.

The progress of city life, the creation of numerous towns, some of them populous capitals, like Alexandria in Egypt, Antioch in Syria, and Seleuceia on the Tigris, the diffusion of Hellenic manners in the East, and the tendency of the Greeks to look further afield—all these circumstances could not fail to give a great stimulus to retail trade. Recent excavations on Delos and on the site of Priene in Asia Minor, north of Miletos, have made it possible to reconstruct at least the material appearance, the setting of this local business. "Let us visit Delos. The streets are lined with shops, most of them quite small; on their fronts are the signs and symbols which advertise their wares; inside, the walls are full of niches. From the objects found on the spot we identify pottery-merchants, ironmongers, sellers of household articles, the ivory-turner, and the sculptor. Near the harbour the shops are grouped according to their special line. . . . Let us go on to Priene. The nearer we come to the markets, the more shops and little windowless workshops there are. Here is a meeting-place of streets—the Small Market. On all sides we see the bakers' shops. The marble tables with the water-channels along them are the butchers' and fishmongers' stalls. Further on is the square of the Great Market; there is a large altar in the middle, and the four sides are lined with spacious arcades, with rows of shops inside."[1]

If we allow for differences due to local conditions, it must have been much the same in all the towns of Asia Minor, both the old Greek colonies of Miletos, Ephesos, Smyrna, Cyzicos, etc., and the new cities founded by the Successors, the various Laodiceias, Apameias, Seleuceias, and Antiochs. We may well imagine the business quarters of Antioch in Syria and Alexandria in Egypt as looking like the *sukh* of a big Levantine port of our own time. There men must have sold, not only the food and other things needed for daily life, but a variety of objects, many of them rare and exotic, intended to satisfy the tastes of well-to-do customers or to tickle the fancy of passing travellers.

The sanctuaries were sometimes also entrepôts and markets. We have seen above[2] that Pessinus was one of the chief trading centres of Galatia and that it owed its prosperity to the famous sanctuary of Cybele. The Egyptian temples were surrounded by shops and hostelries.[3] On the days of pilgrimages and great religious festivals, when pilgrims came pouring in, who had to be fed and wanted to take some memento, talisman, or fetish away with them, this particular form of local retail trade became unusually active.

But, whatever may have been the difference between the true Greek period and the Hellenistic period in respect of the intensity of this local business, it was chiefly in operations on a big scale, export, import, and transit, that the business of the Hellenic world was transformed, or rather took on a new life.

On the Eastern Mediterranean, an active movement of goods—foodstuffs, raw materials, manufactures—flowed from the coasts of Egypt and Asia to European Greece and the West. Some of these goods were produced in the countries which sent them; others had come from further away, and only passed through. Alexandria exported corn, textiles, and paper produced by Egyptian agriculture and industry; it also sent the spices of Arabia and India all over the Mediterranean world. The Phœnician ports distributed purple dyed textiles, Sidonian glassware, and timber from Lebanon

[1] **XLIV**, English pp. 302-03. [2] P. 153.
[3] **LIII**, vol. iii, p. 206.

in all directions; and from their harbours, as from those of Pierian Seleuceia, ships sailed laden with all the goods which the caravans had brought from the countries of the Euphrates and Tigris, Susiana, Persia, and the depths of Asia—perfumes, ivory, precious stones, and silk tissue. From Cyprus copper and wood went to distant markets. Asia Minor supplied the countries of the Mediterranean seaboard with celebrated wools, the saffron and timber of Cilicia, the famous wines, parchment, bronzes, and stuffs, woven or plated with gold, of Pergamon, and the iron of the country of the Chalybes; it also sent, like Syria, everything which came down to the ports of Ionia, especially Ephesos, by the great road which ran from India and Bactriana by the Caspian Gates and Samosata. From the Euxine came livestock, salted goods, honey, wax, furs, and corn.[1] Colchis exported linens which competed with those of Egypt.[2]

It was not only with the Mediterranean and the West that the Hellenistic world maintained active trade relations. The merchants and shipowners of the Hellenized East did not wait for the Asiatic and African goods which they consigned from Alexandria, Seleuceia, Ephesos, Tanaïs, and Byzantion to be brought from the producing country by native shippers; they went to get them by land and by sea, and in exchange they introduced goods produced in the Mediterranean regions. Thus it was that Greek merchants went to India, either by the route along the northern slopes of the Iranian plateau or by the Persian Gulf (or the Red Sea) and the Indian Ocean. Others worked the ports of Arabia Felix and explored the coast of East Africa in the neighbourhood of Spice Cape (Guardafui). In one region or the other they obtained spices, perfumes, ivory, tortoise-shell, precious woods, pearls, and silk; they took to them wheat, wine, textiles, metals, weapons, glassware, and slaves. They also left a great number of gold and silver coins behind them,[3] for they bought far more than they sold on all these distant markets. Strabo even declares that the people of Arabia Felix were said to have amassed enormous wealth

[1] Polyb., iv, 38. [2] Strabo, xi, 2, 17.

[3] Most of these details are supplied by the *Periplus of the Erythræan Sea* (in *Geogr. Græci Minores*, vol. i, pp. 257 *ff.*); this *Periplus* is doubtless not earlier than the first century after Christ, but the trade to which it bears witness was practised as early as the Hellenistic period.

by exchanging their perfumes and precious stones for the gold and silver of other nations and never spending or exporting any of it.[1]

In addition, at the meeting-points of the Hellenistic world and the regions outside, certain peoples acted as middlemen. Such were the nomads of Europe and Asia[2] who met the Greeks of the Cimmerian Bosphorus on the market of Tanaïs; the Armenians who traded with the Scythians beyond the Caucasus, with Central Asia, India, and China in the east, and with Babylon and Cappadocia in the south and west;[3] or, again, the Arabian Scenitæ who controlled the desert country stretching from the Euphrates to Palestine and Syria;[4] or, yet again, the Nabatæans who dwelt in the wilderness near the east shore of the Red Sea.[5]

In respect of trade, therefore, the Hellenistic world must not be regarded as a limited, independent, self-sufficient area. It formed a vast region which faced two ways. It was not a barrier between the Mediterranean basin and the distant and formerly almost unknown and unexplored countries of Ethiopia, Arabia, India, and Central Asia. On the contrary, it maintained communications and economic relations between those two vast zones of different and complementary production. Transit had quite as big a place in economic life as importation and exportation.

The rulers of the kingdoms which sprang from the dissolution of Alexander's empire and a number of Greek cities perceived the profit to be obtained from this situation. No doubt, before the conquest of Asia, customs duties had already been established by certain cities in Greece Proper;[6] but no one had anywhere conceived a complete system of fiscal measures to compare with that with which, for instance, the trade of Egypt was burdened. "There was a net of customs posts," writes Bouché-Leclercq, "drawn tight round all the frontiers of Egypt."[7] Strabo had already remarked on the working of these Egyptian custom-houses.

[1] xvi, 4, 22. [2] Strabo, xi, 2, 3.
[3] **XXXIV**, *Prov.*, English vol. ii, p. 20.
[4] Strabo, xvi, 1, 27. [5] Id., xvi, 4, 23 *ff*.
[6] **XLIV**, English pp. 298-300.
[7] **LIII**, vol. iii, p. 320.

"The most valuable consignments come from these countries" (India and Ethiopia) "to Egypt, and are sent on from there to the rest of the world, so that Egypt gets double duties from them, on coming in and on going out; and the duty is heavy in proportion to the value of the goods."[1]

On the Red Sea, the Egyptian customs no doubt operated in the ports of Berenice and Myos Hormos; in the south, on the Nile, there was a post at Syene; in the north, in the Delta, the most important custom-house was of course at Alexandria, and there were others on the various mouths of the river, at Pelusion, Sebennytos, and Naucratis. The duty on all imports was 25 per cent. of the value; we do not know the export tariff.[2]

This duty on imports and exports was not the only burden laid on goods transported through Egypt. They were subject to a fee for the right to be carried on the roads and canals, and travellers had to pay a similar tax.[3] We know of two toll-houses on the Nile for the barges using the river, at Hermopolis, near the administrative boundary between the Thebaïd and the Heptanomis, and at Schedia near Alexandria.[4] At the gates of towns, and even of villages, goods had to pay duty on going in or out, comparable to the French *octroi*. At Syene, at Thebes, and at the small town of Socnopæu Nesos in the Fayum the duty on outgoing goods was 2 per cent.[5] At Syene and Memphis there was a harbour or wharf duty.[6] Local trade was burdened just as much as big trade. It is not certain that merchants had to take out a trading-licence, but at least they paid for the right to a place on the market.[7] They were also subject to a tax on sales, which seems to have amounted to a fiftieth.[8]

In Egypt, then, there was not a single stage on the way by which goods went either from a harbour of import to a harbour of export or from a place of production to a place of consumption, where the exchequer did not step in and seize some proportion of the value of the goods.

The interference of the Lagid sovereigns also took another

1 xvii, 1, 13.
2 **LIII**, vol. iii, pp. 321 *ff.*; **LXI**, pp. 117 *ff.*
3 **LIII**, vol. iii, pp. 324 *ff.*; **LXI**, pp. 122 *ff.*
4 Strabo, xvii, 1, 16 and 41; **LIII**, vol. iii, p. 323.
5 *Ibid.*, pp. 323 *ff.*; **LXI**, pp. 121 *ff.*
6 **LIII**, *loc. cit.*
7 **LXI**, p. 108.
8 *Ibid.*, p. 124; *cf.* **LIII**, vol. iii, p. 331.

form, which is perhaps more characteristic—the monopoly of certain sales, the monopoly of importation and exportation of certain products. As far as we can tell from the documents, this twofold monopoly applied to salt, nitre or natron, precious stones and ivory, perfumes and spices, wine, oil, papyrus, textiles, and perhaps wheat.[1]

This economic organization of Egypt cannot be explained solely by economic causes or by financial necessities. It is hard to suppose that any protectionist notion underlay the creation of import and export duties on things which Egypt did not itself produce, such as ivory, perfumes, and spices. On the other hand, the rate of these duties would be quite exorbitant, if the Egyptian government had only wanted to recover the costs of the original establishment and the upkeep of the harbours, roads, and canals. The chief, underlying reason of the whole system must be sought in the character of the Ptolemaic kingship, as the heir of the monarchy of the Pharaohs. The sovereign owned all the resources of the country; just as he was, in theory, the sole owner of the soil and absolute master of the human beings who lived on it, so he had the free disposal of everything manufactured by the labour of his subjects, and every commercial transaction—all importation, all exportation, all transit, every sale inside and outside Egypt—was his affair, and the profits should be paid into his treasury. This is another form of the State control of economic matters which we have already seen in the case of agriculture and industry. The roots of that system lay deep in the past of the ancient East, and its tyranny was maintained the more strictly because it was supposed to be justified by the divine nature of the King. The commercial taxation of Egypt seems to me to have the same origin as the land-tax and the organization of agricultural property, the monopolies of manufacture, the taxes on the trades, and the detailed regulations governing industry. Monopolies, taxes, licences, and the rest were simply the various institutions expressing and enforcing the eminent ownership which the King claimed over all nature and all human activity.

While we have fairly detailed and accurate knowledge of the conditions to which trade was subject in Egypt, we have

[1] **LIII**, vol. iii, pp. 238 *ff.*; **LXI**, pp. 60 *ff.*, 120.

far less information about its organization in the other Hellenistic kingdoms—Syria, Pergamon, Bithynia, Pontos, Cappadocia, Macedonia. We know that in many Greek cities and districts there were customs duties, tolls on traffic, and taxes on certain sales,[1] but as a rule we know nothing of their true character. Nevertheless, some very interesting details are given by the ancient authors, by Polybios in particular, about the duties set up by Rhodes and Byzantion.

Rhodes, as we have seen, had become far the most important port of call between the kingdoms of the Orient and the old Greece. Ships from Alexandria, the Phœnician and Syrian ports, and Cyprus stopped there. On every vessel entering their harbour the Rhodians imposed a tax which Polybios calls *ellimenion*, which was doubtless a harbour-due. This tax yielded as much as a million drachmas (rather less than £40,000). When the Romans began to take a predominant place in the Eastern Mediterranean, in 166, after the battle of Pydna, they gave the island of Delos to the Athenians, on condition that it should be a free port. From then onwards the port of Rhodes was abandoned, the yield of the harbour-tax fell to 150,000 drachmas, and the Rhodians sent an embassy to Rome to complain to the Senate.[2] This Rhodian tax was neither a customs duty nor an *octroi*. It was not laid on goods entering or leaving the town or island, but on ships which called at the port. It represented, in Polybios's words, the revenue of the port, ἡ τοῦ λιμένος πρόσοδος. In setting up this tax, the people of Rhodes had acted as a landlord trying to get a profit from his estate. The *ellimenion* was based on the eminent right of ownership which the Rhodians claimed over the actual water of their harbour. Goods which came on to that water, not to be unloaded but merely to remain on it a few days or even a few hours, were taxed, just as the goods which went across Egypt from Myos Hormos, Berenice, or Coptos to Alexandria had to pay import duties, traffic tolls, and export duties, in virtue of a wholly Oriental conception of the sovereignty of the State in every matter and over every thing. The delegate who was sent to Rome by the Rhodians made a point of the fact that his fellow-countrymen had

[1] C. Lécrivain, in **XVII**, *s.v.* "Prosodoi."
[2] Polyb., xxxi, 7, 10 *ff*.

lost the liberty which they had formerly had of themselves settling questions regarding their port.[1]

The case of Byzantion is no less instructive. About 220-219 B.C., being compelled to pay heavy tribute either to the Gauls or to the Thracians, after vainly calling on the other Greek states for help and support, the Byzantines instituted a toll on all ships going from the Propontis to the Euxine or *vice versa.* In this way they hoped to obtain resources with which to pay the tribute. The shipowners and merchants who traded with the countries on the Euxine protested against the tax. Rhodes took charge of the movement. When Byzantion refused to give in, there was a war, in which, in addition to Rhodes and Byzantion, Prusias, King of Bithynia, certain Thracian contingents, and the pretender Achæos, who had rebelled against the King of Syria, played a greater or less part. Byzantion was defeated, and had to abolish the tax.[2] What right could she invoke in favour of its institution ? It was not the consequence or expression of a legal sovereignty, like that of Ptolemy as successor of the Pharaohs in Egypt or that of the Rhodian people over its harbour, which was regarded as an integral part of the domain of the State. The mastery exercised by Byzantion over shipping on the Bosphorus was simply a fact, which Polybios recognizes and describes in forcibly definite language: " Byzantion so controls the mouth of the Euxine (*στόμα τοῦ Πόντου*) that no merchant can enter or leave that sea without her leave." Further on, the historian adds that trade with the Euxine would have been almost impossible if Byzantion had joined forces either with the Gauls or with the Thracians, for, he says, the Bosphorus being so narrow and the barbarians so near, the Euxine would have been unapproachable to the Greeks.[3] Whether Byzantion had a sovereignty by right or a mere mastery in fact, it is none the less clear that the reason for the taxes levied on various commercial operations in the Hellenistic period were much more of a political than of an economic or financial nature. By the collection of these taxes the notion of the sovereign State, that legacy of the ancient East, was expressed and manifested at every moment.

[1] Polyb., xxxi, 7, 10. [2] Id., iv, 46 *ff.* [3] Id., iv, 38, 2 *ff.*

Such a vast and complex system of commercial exchanges, on the top of which there was in the monarchical states and in many cities a fiscal control which was often meticulous and exorbitant, could not have developed and lasted if purchases, sales, and payments had not been conducted by simple, easy methods.

No doubt, in many parts of the East, and particularly in Egypt, debtors continued to make payments in kind, which were accepted by their creditors, and by the State in particular.[1] Perhaps, too, in dealings with the peoples of India, Central Asia, and Arabia, with the nomadic Sarmatians who ranged the steppes north of the Caucasus, and with the tribes which dwelt on the east coast of Africa, simple barter was often practised. The Sabæans, who accumulated masses of gold and silver coin without buying anything from their customers, were certainly an exception. Nevertheless, the practice of settling accounts in hard cash or by banking operations became general in the Hellenistic world.

This increasing supremacy of monetary economy over natural economy was encouraged by various causes, the most important of which were the great increase of the amount of precious metal given over to minting and the institution, if not of a single currency, at least of a number of currencies so closely related that one could pass from one to the other without serious inconvenience.

In Greece, the gold and silver mines of Siphnos were already exhausted, and the mines of argentiferous lead at the Laureion now yielded very little; but the veins of precious metal on Mount Pangæos, on the borders of Macedon and Thrace, which even in the reign of Philip, Alexander's father, had produced a thousand talents a year, made it possible for the Kings of Macedon to issue large quantities of coin. But the great difference was made by the treasures of the Achæmenids, which Alexander seized at Susa, Persepolis, and Ecbatana, where there were enormous masses of gold and silver bullion.[2]

Polybios gives us a very curious piece of information on the subject. In his description of the royal palace at Ecbatana, he tells us that the beams and panelling and the

[1] **XLIV**, English pp. 327 *ff.*; **LIII**, vol. iii, p. 184.
[2] See above, pp. 90-1.

columns of the porticoes and peristyles were plated with gold and silver and that all the tiles were of silver.

"Most of these plates and tiles were removed at the time of the invasion of Alexander and the Macedonians, and the rest when Antigonos and Seleucos seized the power. In the reign of Antiochos, the gold was still on the columns of the temple of the goddess Æna, and silver tiles were heaped up in it in great numbers, and it still contained a few blocks of gold and many of silver. All these riches were used for making coins, on which the royal effigy is stamped, to a value of about four thousand talents."[1]

Later, as trade became intense, the Seleucids, Attalids, and Kings of Pontos and Bithynia were able to obtain gold from Colchis, the Caucasus, the Ural country, Armenia, Bactriana, India, and Arabia.[2] In Egypt itself, the Ptolemies worked rich gold-mines situated in the Arabian Desert.[3] Apart from gold and silver, the mines of Cyprus yielded copper, and trade with the Far East or parts of Iran supplied kings and cities with tin, the two metals needed for making bronze coins. The circulation of metal became general.

That it might not encounter almost insurmountable obstacles, it was necessary that the coins struck by the various states and cities should not belong to systems which were too different from one another, at least as far as gold and silver were concerned, copper and bronze coinages being usually current only in the country where they were issued.

At the time when Alexander conquered the East, the monetary standard adopted by most Greek cities was the Attic standard, the drachma of 4·25 grammes, and the most widely distributed coin was the Attic silver tetradrachm, or four-drachma piece. Alexander made his monetary system fit the Attic standard. Tetradrachms of Attic weight with the effigy of Alexander were struck in most Greek cities down to the Roman conquest in Europe and in Asia.[4] The Attic standard was likewise adopted by the Seleucids.

When Rhodes obtained the commercial supremacy which we have already noted, she caused a system based on a rather different standard to be accepted—that of the so-called Asiatic drachma, doubtless invented in Asia Minor, weighing 3·25

[1] x, 27, 10 *ff*.

[2] Ardaillon, in **XVII**, *s.v.* "Metalla," p. 1851.

[3] Diod. Sic., iii, 12 *ff*.

[4] **XL**, p. 101; F. Lenormant, in **XVII**, *s.v.* "Alexandrei," p. 181; *s.v.* "Drachma," p. 399; **XLIV**, English pp. 325 *ff*.

grammes. The drachmas, didrachms, and tetradrachms of the Rhodian system circulated for about two centuries from East to West. On the obverse they bore the radiate head of Helios, the great god of the island and city of Rhodes, and on the reverse a rose (*rhodon*), the symbol of the state.[1] To the same system belonged the famous *cistophori*, which were so popular all over Asia Minor from the beginning of the second century B.C.[2] These coins were tetradrachms, and got their name from the Dionysiac chest (*kiste*) which adorned the obverse.[3]

In Egypt, when the Ptolemies organized the first monetary system which the country had known and when they substituted their own effigies for that of Alexander, they adopted neither the Attic nor the "Asiatic" standard, but the Phœnician standard, the silver drachma of 3·5 or 3·54 grammes. In this system the silver tetradrachm was the coin in most general use.

There were, then, no fundamental differences between the three or four chief monetary systems used in the Hellenistic East. Attic tetradrachms stamped with the owl, tetradrachms with the image of Zeus Aëtophoros on the reverse, Rhodian tetradrachms, *cistophori* of Asia Minor, and Egyptian tetradrachms represented fairly similar weights of metal and values. While we may grant that a man would not take one of these coins instead of another, the *trapezitai*, or money-changers, had not to make very involved calculations for their operations. Twenty Rhodian tetradrachms or twenty *cistophori* were almost exactly equal to fifteen Attic tetradrachms, and twenty Egyptian tetradrachms to sixteen Attic tetradrachms. That was no serious obstacle to commercial transactions.

Some men seem to have gone further, and to have dreamed of creating a true unity of weights, measures, and coinage. The only instance which we know of an attempt of the kind referred, it is true, to a very limited area, but it is significant all the same. Polybios relates that about the year 220 the Achæans succeeded in establishing this unity in the Peloponnese.[4]

[1] F. Lenormant, in **XVII**, *s.v.* "Drachma Rhodia," pp. 402 *ff.*
[2] **XL**, p. 115.
[3] F. Lenormant, in **XVII**, *s.v.* "Cistophori," p. 1211.
[4] ii, 37, 9 *ff.*

When business developed and trade was done or could be done between regions as far apart as the Euxine, the ports of Syria and Phœnicia, Alexandria, and Great Greece, a comparatively single or similar coinage did not suffice to solve the problem of exchanges. The organization of credit, the action of banks, and the circulation of paper money, which had already appeared in Greek business before the time of Alexander, spread and developed. In many Greek cities, especially in the islands of the Ægean, in Asia Minor, and in Egypt, there were public banks and private banks, which received deposits, usually in specie but sometimes also (as in Egypt) in kind, established veritable current accounts for their clients, issued letters of credit, and effected transfers from one account to another. The use of the cheque became more and more general. Every bank had correspondents in other cities and other countries. All the credit transactions effected by the banks gave a great impetus to commercial operations. There was constant action and reaction; the progress of trade led to the development of credit, and the development of credit contributed effectively to the progress of trade.

Other features characteristic of big business began to take shape. Commercial companies were founded; merchants or banks combined for commercial or financial undertakings which required much capital. In one place the law of competition was evaded by the coalition of the producers or sellers; in another, to raise the price of a commodity, the papyrus of Egypt or the balsam of Judæa,[1] production was limited; elsewhere again, regular corners were effected, which gave a single man the control of the market and a practical monopoly of the sale of the goods in question.[2]

So, during the Hellenistic period, as a result of the work accomplished by Alexander and his successors, economic activity became a very big thing in the Eastern Mediterranean.

The area over which it was practised was very great; it grew up chiefly in Asia and parts of East Africa. Egypt, Syria, Chaldæa, and certain valleys in Asia Minor were

[1] Strabo, xvii, 1, 15.

[2] **XLIV**, English p. 364.

among the most fertile districts in the ancient world. Raw materials, old ones coming in greater quantities or entirely new ones, contributed to the rise of manufactures. Trade, both between the various regions of the Hellenized East and between that East and distant parts of Asia, Africa, and Western Europe, enjoyed a geographical extension and an intensity which it had not known before.

Before Alexander's expedition, the Eastern Mediterranean had been for Greek economic enterprise a blind alley, jealously guarded by the Egyptians, Phœnicians, and Persians. After the expedition, that enterprise found the road lying wide open to the markets of Central Asia, India, Arabia, and the east coast of Africa.

In that greatly enlarged field of action, the Greek genius was able to display its intelligent activity and the spirit of initiative which had long ago been awakened by political liberty and was guided by what was now old experience. The contribution of the East to the organization of the new economic life seems to have been the action and share which the kings claimed in all kinds of production and commerce. The Lagid like the Pharaoh and the Seleucid and Attalid like the Achæmenid were in theory the lords of everything, living and inanimate, in their kingdom. The earth, with its harvests, its trees, its grass-land, and its mineral wealth, belonged to the King, and it was for him that men laboured, manufactured, shipped, and sold. In proportions which varied from one country to another, the King kept for his own immediate profit a part of the soil, industry, and traffic; over the remainder, which he consented to leave to his subjects, his eminent right of ownership was expressed by duties and taxes of various kinds, many of them very burdensome. It was in the Oriental monarchies that economic nationalization had its birth, as a logical consequence of the divine nature ascribed to the King.

While all round the Eastern Mediterranean and in the regions of Asia nearest to that sea, this new economic life was developing, having at once something of the old methods practised in independent Greece and something of those characteristic of the states of the ancient East, the West, after long remaining barbarous or isolated, was opening more and more to influences of every kind from the East. Its economic life was developing. After first being subjected to the influence of Greece and the East, presently, as the political power of Rome dominated and then absorbed the other states, it would contribute to the establishment of the unity of the Mediterranean world in the special domain of economic facts.

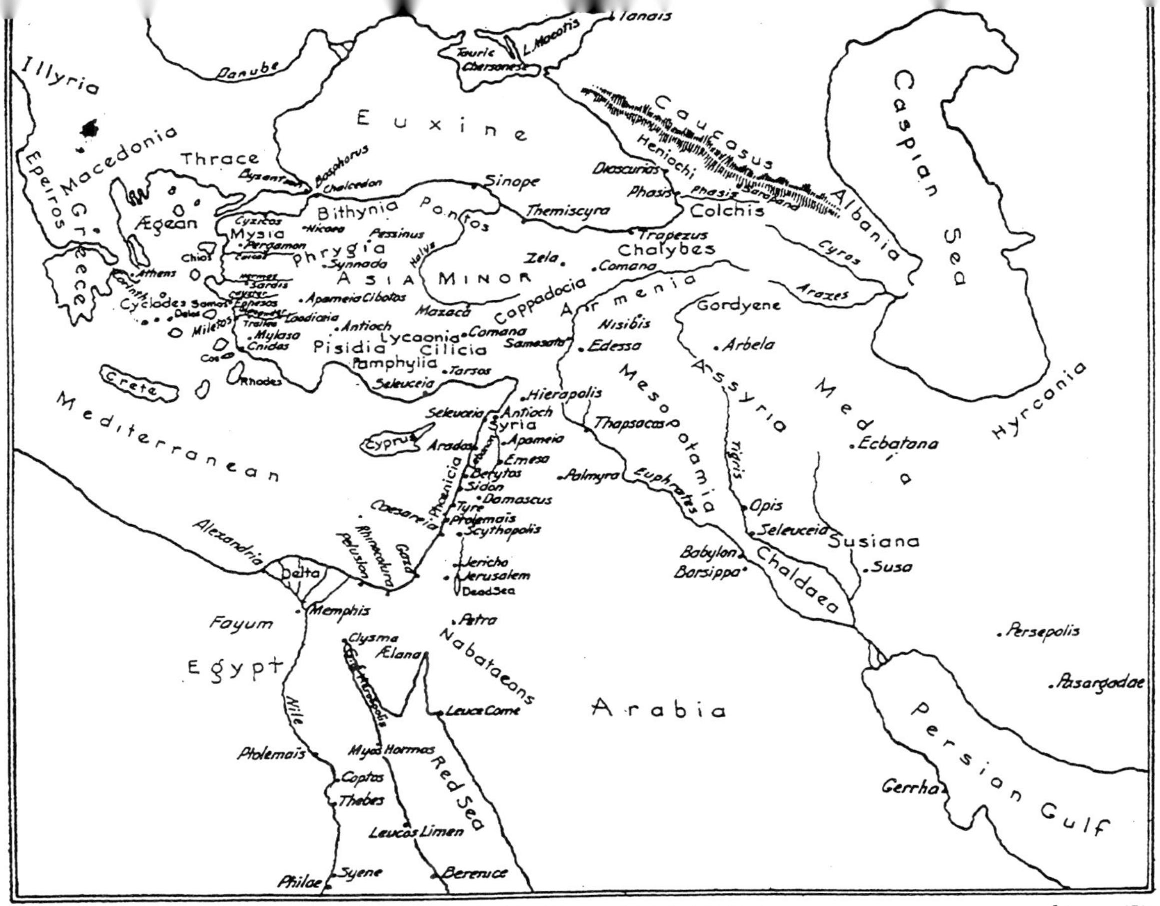

MAP III. THE HELLENISTIC WORLD.

[face p. 170.

PART III

THE ECONOMIC LIFE OF THE WESTERN MEDITERRANEAN AND THE NEIGHBOURING REGIONS TO THE ESTABLISHMENT OF THE ROMAN EMPIRE UNDER AUGUSTUS

CHAPTER I

THE ECONOMIC LIFE OF THE BARBARIAN WEST: NORTH AFRICA, THE IBERIAN PENINSULA, GAUL. THE PROGRESSIVE INFLUENCE OF CARTHAGE, GREECE, AND ROME

In the course of the thousand years before the Christian era, economic life in the Eastern Mediterranean went through brilliant and fruitful stages, both before and after the conquest of the East by Alexander. Already organized in a collective form in the society depicted in the *Odyssey*, the *Iliad*, and the poems of Hesiod, it advanced from stage to stage, as the field of action of Greek civilization extended, until it acquired a remarkable amplitude, variety, and complexity. The monarchical conception of the State inherited from the old Eastern empires had a very special influence on the economic organization of the Hellenistic period, and that influence continued to be felt for many centuries.

The regions surrounding the Western Mediterranean present a different spectacle during the same period. No doubt, at certain points of that basin, the influence of the East and Greece contributed to the development of economic life. The presence of certain Greek colonies like Marseilles, the power of the Carthaginian Empire in North Africa and the Iberian Peninsula, and the brilliance of the Etruscan civilization do not affect only the history of Greece, Carthage, and Etruria; they have played an essential part in the history of the whole West. Nevertheless, almost down to the

Christian era, this history of the West almost belongs to prehistory. Here one can follow, better than on the banks of the Nile or the Euphrates and better than on the shores and islands of the Ægean, the successive stages, rapid and slow, of economic progress from the earliest Stone Age to the time when these countries, long isolated and little known, were brought by the Roman conquest into the main stream of the economic life and civilization of antiquity.

I

The Economic Life of Palæolithic Man

In Palæolithic times, when his dwelling was a cave or a shelter under a rock, man lived entirely by hunting, fishing, and gathering a few wild plants or fruits. He did not yet practise agriculture or stock-breeding.

But industry already existed. It consisted chiefly in chipping flint, working in bone and ivory, and preparing the hides of beasts. From the primitive axe known as the *coup de poing* of the Chellean epoch to the Solutrean blades and spear-heads shaped like willow and laurel leaves, the working of flint and other hard stones was perfected. This industry supplied man, not only with the weapons needed for fighting wild beasts, but with several implements—knives, punches or drills, engraving-tools, scrapers. From hard materials of animal origin—bone, ivory from the tusks of mammoths and elephants, reindeer-horn—he trained himself to make a number of articles, the use of which is not always easy to determine—harpoons, assegai-heads, needles with eyes, pins, spatulas, spear-throwers, *bâtons de commandement*, and perhaps even ornaments. There was no spinning or weaving, but man could prepare the skins of beasts for use as clothing, and it has also been supposed that he made leather vessels. He had observed, lastly, that certain natural substances gave things a colour, and used them to tattoo himself. Pottery, even of the clumsiest kind, seems to have been almost unknown; at least, it has not been found on the Palæolithic sites, except at a few stations in Belgium.[1]

[1] **XIX**, vol. i, pp. 169 *ff.*; **LXXVIII**, English p. 66.

How was this primitive industry organized? We do not know. It seems likely that every man, every head of a family, was able to chip flint, carve bone, ivory, and stag or reindeer horn, and prepare the skins of beasts. We do not find in the distribution of the Palæolithic dwelling-places anything which suggests the specialization of industry. The various stations have all yielded products of this industry in similar form and in similar proportions. A genuine workshop is, however, reported at St. Acheul in the French Department of the Somme.[1] On the other hand, the works of art of the Quaternary period—cave-paintings, bone and ivory figurines, engravings on bone and reindeer-horn—testify to the technical skill and sense of life and style reached by their creators. Can we admit that, in a world in which art was so developed, economic activity was confined to the individual manufacture of weapons and tools as they were needed, from one day to another? It is wisest not to give a definite answer to the question.

Also, did the tribes of Palæolithic times have any knowledge of trade? "On several occasions," Monsieur J. Déchelette writes, "stations of the Reindeer Age have yielded pierced shells, used as ornaments, which are proved by their scientific class to have come from distant places." So, too, "some Quaternary flints in the Belgian stations have been observed to be of foreign origin." But, the learned archæologist adds after mentioning several facts of the same kind, it is not possible "to determine for certain whether this transport . . . was due to a true trade or merely to the many migrations of the Quaternary tribes."[2] Monsieur J. de Morgan, on his side, while admitting the existence of commercial relations as early as the Palæolithic period, of exchanges carried out between clan and clan, between tribe and tribe, does not believe that any trace has been left of these operations. With reference to the discovery in caves in the centre of France of shells from the Ocean and the Mediterranean, he asks whether they should not be explained as spoils taken from conquered enemies.[3]

So, according to recent observations, the economic activity

[1] *Ibid.*, English p. 41.
[2] **XIX**, vol. i, pp. 175 *ff.*
[3] **LXXVIII**, English p. 269.

of Palæolithic man may be defined as follows: no agriculture or stock-breeding; rudimentary industry; very doubtful commercial exchanges.

II

The Economic Life of Neolithic Man

The Neolithic Age, which, according to the specialists, is chiefly distinguished, in respect of stone implements, by the substitution of polished stone for chipped or simply flaked stone, made one of the most important advances in the economic evolution of mankind.

Agriculture made its appearance. Man no longer relied for his food on hunting, fishing, and gathering wild plants and fruit. He knew and doubtless cultivated corn—wheat, barley, millet, rye, oats. Fruit had a big place in his diet; nuts, sloes, and strawberries were doubtless provided by nature spontaneously, but apples, pears, and grapes seem to point to artificial growing.[1] The tilling of the soil had come into being, though under what influences we cannot determine or even guess.

The domestication of animals, and consequently stock-breeding, developed at the same time. The dog entered upon his rôle of faithful comrade, helper in the chase, and watchful guardian. Man began to make use of the strength and speed of the horse. The bovine, ovine, and porcine races supplied him with meat, milk, leather, and wool, and perhaps oxen already served as beasts of burden.[2]

Closer bonds than before were established between man and the soil on which he lived, and from which he obtained at least part of his own food and that of the beasts which he collected about him. Caves and rock shelters were abandoned for dwellings built in the open, artificial constructions placed together in what were often fairly big agglomerations. On the dry land, they were round or rectangular huts with a hole dug below the floor level in the middle to hold the hearth and walls made of wicker hurdles covered with clay. The villages formed by these huts were often built on promontories or spurs in a naturally strong position; elsewhere they were surrounded by defence-works,

[1] **XIX**, vol. i, pp. 342 *ff.* [2] *Ibid.*, pp. 337 *ff.*

regular ramparts. It was not only on dry land that human habitations arose and collected. For better protection against wild beasts, which were still a danger, or against hostile tribes and pillaging neighbours, men settled in lakes, on piles. This was the age of the lake villages or palafittes. Even the dwellings of the dead underwent the same evolution; funerary megaliths, galleries (*allées couvertes*), and dolmens constituted veritable necropoles, cities of the dead. There is no doubt that society developed in the Neolithic Age, that men became more and more accustomed to come together, and that economic life was thereby affected.

The birth of agriculture and stock-breeding and the progress—relative, no doubt, but real and significant—of comfort in domestic and in collective life led to the birth and development of several industries.

Stone, wood, bone, and horn served for the manufacture of all kinds of things, weapons, tools, utensils, and pendants and other articles of adornment. The polishing of stone, the cutting of wood, and the working of bone and horn furnished Neolithic man with the materials and implements needed for an existence which had become almost sedentary and industrious—axes of many shapes, clubs, adzes, gouges, chisels; the piles of their platforms, the still rudimentary roof-timbers and floor-boards of their dwellings, the handles of their tools and weapons, even dug-out canoes and a few bowls; daggers, drills and punches, arrow-heads, javelin-heads, pins, needles, and crude trinkets; picks for use in the flint-beds, barbed harpoons, etc. The mere list of these articles, short as it is, shows the great importance assumed by peaceful and productive occupations by the side of hunting, fishing, and fighting. The character and horizon of life were transformed.

The ceramic and textile industries began to play a big part in economic life.

The earthenware vases at that time were made " of ill-refined clay mixed with grains of quartz, which perhaps served to give it more consistency. The colour is rarely even all over the surface of the vase. At breaks, the clay, which the imperfect firing has not penetrated equally all through, often presents a more or less reddish colour on the two faces and a grey colour inside. . . . The vases were fired in the open air, not in a kiln. . . . The use of the wheel being unknown, the potters made all their goods by hand."[1] It is not known whether this industry was specialized or domestic at the time. The primitive nature of the plant and methods of manufacture, the crude simplicity of the ornament, which usually consisted of lines or dots incised to various depths, and the clumsiness of most of the shapes given to these vases do not suggest a highly developed technique. It is very likely that the vases in common use were made in each village, perhaps in each family. In spite of many recent discoveries and discussions, there is still much obscurity in the history of Neolithic pottery.

The textile industry was no more advanced than the ceramic. Flax and wool were spun and woven. Remnants of tissues, ropes, and fishing-nets have been found in lacustrian stations in Switzerland, besides several of the implements used, such as whorls and bobbins of terra-cotta.[2]

Lastly, we should mention—though one cannot describe these occupations as industries—the production of flour by means of stone querns, or rather crushers, and the use of flat polished stones for the crushing of colouring matter, especially red and yellow ochre, intended for painting and tattooing the body.[3]

The evolution which transformed economic life in the Neolithic Age affected the relations of peoples with one another, and therefore trade. In Palæolithic times, men had generally been content with the food and materials which nature provided in the immediate neighbourhood of their dwellings, whether caves, rock shelters, or stations on headlands, flat hills, or alluvial ground beside rivers. Moreover, the hunters, in their wanderings after game, had been able to obtain, sometimes far away from their starting-point, objects which they then carried miles away with them, such as the sea-shells from the Atlantic or the Mediterranean which have been found in various inland stations. It may be said that in that age it was man who moved about and went to the places of production.

With the more settled life which marks the Neolithic Age, with the development of agriculture, stock-breeding, and

[1] **XIX**, vol. i, pp. 545 *ff*. [2] *Ibid.*, pp. 579 *ff*.
[3] *Ibid.*, pp. 565 *ff*.

industry, and with the predominance of peaceful, industrious activity, the opposite movement set in. Henceforth man was more stationary; he attracted to his dwelling, to the centres of consumption, the natural and manufactured products which he needed. The many observations to which the study of the Neolithic deposits has given rise allow one to state that a very big trade distributed various materials of mineral origin over areas some of which were quite extensive. Flints chipped in the workshops of the Grand Pressigny in Touraine were exported to Brittany, the north of Gaul, and even Switzerland.[1] Jade, which was used for making polished axes, is not found in a raw state in Western Europe except in certain parts of Switzerland and Styria; jade axes have been found in Brittany.[2] Obsidian, too, the deposits of which can be placed exactly, in view of its volcanic origin, travelled over a wide area.[3] The trade in amber and callaïs extended still further. Amber, or more exactly yellow amber, comes in Europe from the shores of the Baltic and the North Sea. As early as Neolithic times, it was used for making necklaces, not only in Scandinavia, but in various parts of central and southern Gaul.[4] The question of callaïs is less clear, because we do not know exactly from where it came; it is, however, certain that this substance, which was very rare and has now disappeared altogether, was conveyed to many regions a long way from one another, since beads of it have been found in Morbihan, in Provence, at the foot of the Pyrenees, and in Portugal.[5]

We see, from these examples, which are the best known up to the present, that Neolithic trade went on over big areas, since the flints of the Grand Pressigny spread far to the north-east and to the east, from the banks of the middle Loire to the country of the Scheldt, the Meuse, and the Jura; since the jade from the Eastern Alps reached Brittany; since the amber gathered near the mouths of the Vistula, Oder, and Elbe was already conveyed to the shores of the Atlantic and the Mediterranean.

It is probable that this already active trade travelled overland. " The genius of the Romans did not create the

[1] *Ibid.*, p. 629.
[2] *Ibid.*, pp. 627 *ff.*
[3] *Ibid.*, pp. 628 *ff.*
[4] *Ibid.*, pp. 623 *ff.*
[5] *Ibid.*, pp. 620 *ff.*

road-system of the conquered provinces, lock, stock, and barrel. Rome did not have to pioneer all the great trade-routes of her empire. The monumental causeways which she built of lasting materials took the place of tracks which had long been drawn, in the course of the ages, at the cost of infinite efforts, by the ancient inhabitants of Europe. They were modest tracks, but they had facilitated the diffusion of primitive civilization, and the patient discoveries of prehistory are gradually rediscovering them under their covering of Roman flagstones."[1]

At the same time as these tracks, the rivers were used. One is, at least, entitled to assume this from the fact that Neolithic man used dug-out canoes and that, according to classical tradition, the amber came to the Mediterranean by a river which Herodotos calls the Eridanos. It has been supposed, with justice, that the Vistula and Dniester in the east, the Elbe and Moldau in the middle, and the Rhine and Rhone in the west were the chief routes of the amber trade.[2]

Did navigation by sea yet play any part in commercial relations? To explain the diffusion of amber in Western Europe, M. Déchelette adds to the river-routes mentioned above " the great sea-route by the Atlantic and the Strait of Gibraltar "; he admits, however, that in Neolithic times this route was only of secondary importance.[3] M. de Morgan, on the contrary, believes that men only ventured on the waves to fish. " Their vessels were far too unstable to warrant the risk of long voyages along coasts that were often very inhospitable."[4]

Primitive and rudimentary as the methods, equipment, and organization of economic life still were in those distant days, a very important stage had been passed by man. Agriculture, the domestication and rearing of animals, many industries, and, lastly, commerce had come into being. The essential forms of economy had appeared. They might afterwards develop, become more complex, and be perfected; but they existed. Man had them, knew them, used them.

[1] **XIX**, vol. i, pp. 629 *ff.*
[2] *Ibid.*, p. 626.
[3] *Ibid.*, *loc. cit.*
[4] **LXXVIII**, English p. 274.

III

The Bronze and Iron Ages. The Protohistoric Period

The Neolithic Age is marked by the appearance of agriculture, stock-breeding, and the textile and ceramic industries; the two distinctive features of the following period are the birth of metal-working and the development of commercial relations, both among all the countries of Central and Western Europe and between those countries and the regions washed by the Eastern Mediterranean. Economic life all over the West then took on an activity previously unknown, which would endure and increase from century to century, and would not enter upon a new stage until the time of the Roman Empire.

About agriculture and stock-breeding there is nothing to say, except that its progress is proved by the extension of the use of the sickle, by rock-carvings of ploughmen driving ploughs drawn by two oxen, and by the abundance of woven stuffs, chiefly wool, found, in a marvellous state of preservation, round several skeletons.[1] At the end of the protohistoric period, although huge, dense forests still existed, the greater part of the ground, at least in Gaul and the Peninsula, was covered with various crops and pastures. We know this from various Greek and Roman authors, such as Cæsar, Strabo, and Diodoros.

On the other hand, the use of the metals does not seem to have had any perceptible effect on the general conditions of habitation. Writing of Gaul, M. Déchelette says: "Not only was all attempt at luxury and comfort banished from them, but down to Cæsar's time they were built of light materials—bits of wood, planks, or a mixture of mud and branches, coated over with clay and roofed with thatch."[2] One should, however, mention a few walls of dry-stone combined with wooden beams found, for example, at Bibracte and Alesia.[3] But this was hardly an advance, and M. Déchelette's remarks are not in any way weakened. It is, moreover, probable that they apply to the whole of the West.

[1] **XIX**, vol. ii, 1, pp. 266 *ff.*, 497, 307 *ff.*
[2] *Ibid.*, pp. 111 *ff.*
[3] J. Toutain, in *Pro Alesia*, N.S., vol. xi (1925).

One should remember, too, that during the centuries just before the Christian era big migrations took place north and west of the Alps. Some of these migrations ended in permanent settlements, like that of the Celts in Gaul, Switzerland, and northern Italy; others were temporary invasions, accompanied by looting and destruction, like those of the Cimbri, Teutones, and Suevi. While the settled, industrious life of many tribes encouraged economic progress of every kind, the intermittent or habitual nomadism of others opposed and hampered it. This was the case in Germany, where wandering bands never ceased moving about through forests and swamps,[1] and it was the case in North Africa, except in the parts which came directly under the influence of Carthage.

The textile and ceramic industries continued to advance. The potters learned to use the wheel, and in the shape and decoration of their vases they tried, still very feebly, to imitate the works brought from Crete and Greece by Ægean, Phœnician, and Greek mariners. They emphasized their incised ornament with crude colours. Those of the Iberian Peninsula painted their wares with the brush, covering the belly and neck with geometrical patterns and floral and animal decoration but very rarely attempting the human form.[2]

We have less information about the progress of the textile industries. We do not know at all whether they lost their domestic character and became specialized. They do not seem to have undergone the influence of the East to the same extent as pottery and metal-working. M. Déchelette thinks that the Gauls may perhaps have taken the most characteristic part of their costume, their trousers, from the Scythians and Persians. What is certain, is that the draped garments usual among the Greeks and Romans were not introduced among them until after the Roman conquest.

Whatever may have happened in the ceramic and textile industries, the great economic novelty of the time was metal-working. What were the origins of this craft in the West? The opinion most generally accepted among prehistorians is that the revolution started in the Mediterranean and Asiatic

[1] Cæs., *Gall. War*, vi, 21 *ff.*; E. Babelon, *Le Rhin dans l'histoire*, i, pp. 114 *ff.*

[2] P. Paris, *L'Art et l'industrie de l'Espagne primitive*, vol. ii, pp. 1 *ff.*

East. The two metals earliest known and used were copper and gold. "It would seem to be definitely proved today that the knowledge of copper reached Gaul simultaneously from the South and the East, that it came from the Black Sea and the Ægean, a district where this industry—according to specialists in Ægean questions—began towards the early part of the third millennium before our era; though naturally it took long centuries before it was propagated as far as the British Isles and Scandinavia."[1] It is not impossible that gold, which appears in the Egyptian and Chaldæan tombs at the same time as copper, followed the same route. The knowledge of bronze, an alloy of copper and tin, was likewise transmitted from the Ægean lands to the West.[2] Still later, finally, it was perhaps from Western Asia and the Mediterranean East that iron-working spread to Central, Northern, and Western Europe.

However one may attempt to solve these problems of origin, once the initial impetus was given there is no doubt that the metal industries found a very propitious soil everywhere in the West. The metals were supplied by very numerous and very rich deposits. To mention only the most important, there was copper in the Peninsula, tin in Portugal, Brittany, and Cornwall, iron in Illyria, Istria, Noricum, Lorraine, Burgundy, Berry, the Pyrenæan region, etc., and argentiferous lead in the south of Spain, and there were mines and placers of gold all over Gaul and Ireland. This is not the place to recall in detail all the chronological and economic phases through which the metal industries passed from their appearance in the West down to the Christian era, Special works, particularly Déchelette's *Manuel d'archéologie préhistorique* and de Morgan's *Prehistoric Man*, will give the reader full information about the Bronze Age and its subdivisions, the first Iron age or Hallstatt period, and the second Iron age or La Tène period. Here we have simply to try to see the influence exercised on ancient economic life by the progress of metal-working in the centre and west of Europe.

First of all, the first result of that progress was to place a large quantity of ores at the disposal of man. It was not only the population of the country where the copper, tin,

[1] **LXXVIII,** English p. 105; *cf.* **XIX,** vol. ii, 1, p. 2.
[2] **LXXVIII,** English pp. 114-15; **XIX,** vol. ii, 1, p. 3.

argentiferous lead, iron, or gold was found that benefited by these hitherto unexploited and unproductive riches. At an early date the Phœnicians, and after them the Carthaginians, took advantage of the silver-mines of the Peninsula and the tin-mines dotted along the Atlantic seaboard from Lusitania to Britain. The silver obtained from Iberia was used by the Carthaginians to pay their army of mercenaries. The tin of the Atlantic regions was transported both by sea and across Gaul to the Mediterranean, and brought on Phœnician ships to the great industrial centres of the classical East. Iron spread practically everywhere in the west and north of Europe. This diffusion of metallic raw materials of western or northern origin was an economic phenomenon of great importance.

The discovery and exploitation of these raw materials had as a result the organization and development of new industries almost all over Europe. No doubt, Neolithic man had extracted flint with his rudimentary apparatus, and Déchelette writes, "Our knowledge of Neolithic industry justifies us in saying that the art of mining was earlier than that of metal-working";[1] but that primitive art, practised with tools of stone or staghorn, cannot be compared, even remotely, to the technical methods, already complex and almost scientific, used in mining copper, tin, gold, and silver. Long before the Christian era, miners in the Peninsula, in Gaul, in Britain, and at the foot of the Alps were digging galleries to reach the strata and veins of ore, draining off the water which constantly threatened to flood those galleries, and extracting the precious or base metals from the enclosing gangue of stone by washing and smelting, and had observed that a mixture of various metals in the right proportion produced alloys. It may not be certain that the alloying of arsenic, antimony, and zinc with copper was consciously practised by the ancients; but at least it is undeniable that they had recognized the union of copper with tin, and practised it with great skill. In the Bronze Age, and still more in the Iron Age, foundries and forges became numerous. On the site of these metal-works, crucibles, blast-pipes, ingots, and moulds have been found; and the examination of the articles manufactured has revealed the usual processes of the in-

[1] **XIX**, vol. i, pp. 355 *ff*.

dustry—the use of rivets, and then of solder, to join and fit separate pieces and the invention and progress of engraving, chasing, stamping, repoussé-work, and enamelling.[1]

From foundries and forges there poured forth, in hundreds and in thousands, offensive and defensive weapons and armour, swords, spears, daggers, axes, arrow-heads, javelin-heads, helmets, breast-plates, shields, arm-guards, and chariot wheels and fittings; tools and implements of all kinds, sickles, knives, adzes, chisels, files, saws, punches, hammers, and anvils; harpoons and fish-hooks; many utensils, such as vases of various shapes and sizes, spits and forks for roasting meat, firedogs, ploughshares, and bits; ornaments and toilet articles, bangles, rings, necklaces, pins, fibulas, and combs; fetishes and amulets, the most popular of which seem to have been disks. Bronze, iron, gold, silver, and lead were, simultaneously or successively, used in making all these things.

This metal industry was not at first uniform in all parts of Central, Northern, and Western Europe. At the beginning of the Bronze Age, prehistorians make a distinction in this respect between several provinces. Apart from the Eastern Mediterranean and Italy, they distinguish an Iberian province, a Western province, comprising France, the British Isles, Belgium, Switzerland, and South Germany, a Hungarian province, and a Scandinavian province. Gradually, under the influence of the increasingly active relations which grew up between these regions, differences tended to disappear. Goods manufactured in countries very far apart came together. Huge migrations, like those of the Celts from the Atlantic to the Black Sea, carried types of weapons and tools which were at first only in local use to very distant places. At the same time, the increasing importation of Greek manufactures and works of art had a similar effect on the aforesaid provinces, or at least on most of them, and may have helped to reduce regional differences during the centuries immediately preceding the Christian era.

At that time, as later, the interdependence of trade and industry was one of the characteristic features of economic life. The wealth of Western Europe in ores and metals, the deposits and production of which were fairly definitely localized, gave rise to a very big current of exportation. In

[1] **XIX**, vol. ii, 1, pp. 175 *ff.*

addition to the amber trade and its routes, which had been established in the Neolithic Age and grew steadily more active, there were now the tin, bronze, iron, and salt trades and their routes. The new trade and routes did not only run from north and west to south and east, from Central and Atlantic Europe to the Mediterranean countries; they brought Iberia, Gaul, and the British Isles into touch with Scandinavia and Central Europe. Tin was carried by sea from the shores of the Atlantic both to the Mediterranean and, by the North Sea, to the Baltic; by land, it crossed Gaul from north-west to south-east. Ingots of bronze from the countries rich in copper and tin mines went over the passes of the Alps. It is probable that the iron of Noricum and Styria travelled northwards and westwards as well as southwards. The salt trade had become very important; the general directions which it took were determined by its nature and source, running chiefly from the coasts where it was obtained to the interior.

Like raw materials, manufactured goods travelled over the west and centre of Europe. " One could make up a whole list of the objects of the Bronze Age found in our (Gallic) provinces or in the Swiss lake villages, having been imported there from abroad. I have spoken of the British gold ornaments, the Italic helmets and daggers, the Iberian or Scandinavian war-hammer of Kersoufflet, the gold pin of Serrigny, other pins of the Bronze Age imported into Gaul from Central Europe, and many other objects of foreign origin. A few more discoveries which are no less significant may be mentioned. Archæologists have long recognized the Scandinavian origin of a hanging-vase and the half of a fibula found in 1878 at Corcelette, on the Lake of Neuchâtel. . . . A sword, of a type characteristic of the Hungarian Bronze Age, is said to have been found at Beynost in the French Department of the Ain. Another sword of the same type . . . has been found at La Plaine near Martigny in Switzerland."[1] These currents of trade went on in the Iron Age. " It was in great part to this internal and external trade that the La Tène civilization owed . . . its wide diffusion into regions where the Celts had not penetrated."[2]

Internal and external trade—indeed, it was during the

[1] **XIX**, vol. ii, 1, pp. 395 *ff.* [2] *Ibid.*, 3, p. 1574.

metal age that Western and Central Europe began to entertain frequent and continuous relations with the already brilliant civilizations of the Mediterranean, with Italy and Greece. When, near Montlaurès and Enserune, between Narbonne and Béziers, cemeteries were discovered whose furniture consisted partly of Greek vases adorned with painted figures and Campanian and Italo-Greek wares dating from the sixth century B.C. at the earliest, it was readily believed that these finds revealed the first penetration of a still barbarous Gaul by the products of Hellenic industry and art. M. Piroutet has recently shown that this belief is incorrect, and that long before the sixth century many objects, particularly metal objects, had been imported from the Mediterranean world, not only into the coast-districts close to the Greek colonies, but far into the interior of Europe, all over Gaul and into Switzerland and South Germany.[1] M. Déchelette had, moreover, pointed out that, even in the Bronze Age, pigs and ingots of copper which were certainly of Ægean origin had gone as far as the Rhone and the Danube.[2] In the Iron Age, this trade extended, reaching the west and north of Europe, whither it took, among other goods, bronze vases, superior pottery, coral, and glass trinkets.[3]

This traffic took two main directions. From the top of the Adriatic, where the Romans afterwards founded Aquileia, it went either up the valleys of the Po and its affluents, in particular the Adige and Ticino, or due north up the Tagliamento and Isonzo, until it came to the chief passes of the Alps, the Great St. Bernard and Simplon in the west and the Brenner and Predil in the east. North of the Alps, it spread out like a fan, as it were, towards the middle Rhine, the upper Danube, the Moldau, the Elbe, and the Oder. From the mouths of the Rhone, especially after the foundation of Marseilles, Italo-Greek imports followed the long corridor of the Rhone and Saône and then spread by the Loire and Seine, by the Meuse and Moselle, all over western and northern Gaul. In the Peninsula, roads doubtless ran to the interior from the Greek colonies and the Phœnician and afterwards Carthaginian ports. Emporiæ (Ampurias) at the foot of the Pyrenees and Gades (Cadiz) on the Atlantic must have been

[1] *Congrès de Rhodania* at Avignon (Sept., 1924), pp. 92 *ff.*, no. 954.
[2] **XIX**, vol. ii, 1, p. 400.
[3] *Ibid.*, 3, pp. 1573 *ff.*

ports of distribution for goods coming from the East. To these land-routes one should add the great sea-route which went out of the Mediterranean by the Strait of Gibraltar and then followed the coast of Portugal and western France to Brittany and Britain.

With this development of trade went a remarkable and continuous progress in the organization of exchanges. So far as prehistoric and protohistoric archæology can tell at present, the instruments of exchange used in Western Europe were at first metal objects, single or double bronze axes and then disks; in Britain, iron bars of a fixed weight were used for the same purpose. In the fifth century B.C., money appeared. The Greeks introduced coins which came from their mother-country into Gaul and the Peninsula; the famous treasure of Auriol, found near Marseilles in 1867, contains coins of Phocæa, Lesbos, Ægina, and other Greek cities. Later, Macedonian staters penetrated fairly far into Gaul. The coins issued by Marseilles and by Rhode and Emporiæ in Spain travelled far and wide. It was in imitation of the Greek coins that the Gauls and Iberians started, in the last centuries before the Christian era, to strike native coinages. We need not enter into a discussion of numismatics here. But we should note that the use of money, which became general in Western and Central Europe before the Roman conquest, was taught to the native populations by the Greek traders, and that it was a necessary consequence of the extension of Greek and Italo-Greek commerce in those regions.

So, then, at the time when the various countries of Western Europe were made subject to Rome and combined by her in a single state, they had passed some essential economic stages. Agriculture and stock-breeding were practised almost everywhere. There were many industries, which made use of the abundant raw materials which man was able to extract from the soil or to produce by his labour. An active trade had grown up between the regions of that vast territory, many of them far away from one another, and all, directly or indirectly, had relations with Italy and Greece.

While the barbarous countries of Europe were passing through the successive phases of this economic evolution, two other countries, which had long stood outside Hellenic civilization, were taking a preponderant place in the Western Mediterranean—North Africa and central and northern Italy. Two cities there played the chief rôle—Carthage and Rome—until the day when Rome, having triumphed over Carthage, became the unchallenged mistress, first of the West and then of the whole Mediterranean world.

CHAPTER II

THE ECONOMIC LIFE OF CARTHAGE[1]

THE economic life of North Africa, from the Greater Syrtis to the Atlantic, remained stationary much longer than that of Western Europe. The influence of the Ægean civilization does not seem to have made itself felt here. The pioneers of progress were first the Phœnician mariners and then Carthage.

At the time when the sailors of Tyre and Sidon were establishing their first ports of call on the north coast of Barbary, the peoples then inhabiting the country had not advanced beyond the civilization of the Stone Age, being still at the Neolithic stage. The most ancient metal objects used by them were probably brought from abroad by sea.

Although our information about the beginnings of agriculture in North Africa is very meagre and very vague, we may suppose that, at least in some parts, wheat, barley, and some vegetables, including the bean, were grown before any Phœnician influence appeared. In any case, stock-breeding already had an important place in economic life; oxen, sheep, goats, and horses were domesticated. They had perhaps come from the East, from Egypt.[2]

According to the most probable reckoning, it was in the twelfth century B.C. that the first Phœnician ships appeared in the waters of the Western Mediterranean. They needed ports of call along the African coast, on the way to the Spanish mining districts, and these ports, where the earliest traffic with the natives doubtless began, subsequently became permanent "factories," real colonies. According to tradition, the earliest Phœnician settlements were, from east to west, Leptis Magna (Lebda), Hadrumetum (Susa), Utica, Hippo

[1] For this chapter I have chiefly made use of S. Gsell's masterly *Histoire ancienne de l'Afrique du Nord*, vol. iv, bk. i (**LXXI**); *cf.* **LXXIV**.

[2] **LXXI**, vol. i, p. 234.

Diarrhytos (Bizerta), and Hippo Regius (Bona), and at this time, beyond the Pillars of Hercules, Lixus was founded (near Larash, south of Tangiers). Meanwhile, the Tyrians and Sidonians established themselves on Malta, at the southern point of Sicily, in Sardinia, and at Gades in Spain.

Carthage was founded about two hundred years after these beginnings of Phœnician colonization in the West; one may safely say that it was at the end of the ninth century.[1] She rapidly became the most prosperous and powerful of all these colonies. When Tyre had been first weakened and then ruined by the despotic rule of the Kings of Assyria, Chaldæa, and Persia, Carthage took upon herself the protection of the Phœnicians against their rivals, the Greeks, all over the West, and became the capital of a veritable empire of a marine and commercial kind. At the beginning of the fifth century, her supremacy was recognized, willingly or otherwise, by the settlements which Tyre and Sidon had founded in the Western Mediterranean and along the Atlantic coasts northward and southward from the Pillars of Hercules. Her sway extended over western Sicily, Sardinia, the Balearic Isles, the west coast of Iberia, and all North Africa as far as the west of Cyrenaica.

In Africa, not content with imposing her authority on the coast-towns, she annexed a fairly large territory in the interior. Although the exact boundaries of that territory are not known, historians are agreed that it comprised the north and most of the centre of the present Tunisia. This was the starting-point and base of an influence which cannot be disputed. "Carthage," writes M. Gsell, "already a great Mediterranean port and the capital of a vast sea-empire, now became an African capital as well. She spread her civilization in the country which she annexed, and then beyond her own territory, among her vassals and allies."[2] The influence which she could then exercise was not only political; it contributed to the progressive transformation of the economic life of North Africa.

[1] *Ibid.*, p. 401. [2] *Ibid.*, p. 465.

I

The Practice and Science of Agriculture at Carthage

To whatever extent the earlier populations of North Africa may have grown certain crops and raised livestock, it was only after the constitution of the Carthaginian Empire that the country really became acquainted with agriculture, that the various resources of the soil were exploited scientifically and methodically, that a serious effort was made to cultivate its vegetable and animal wealth. No doubt, that rural economy was only applied to a fairly small area, the territory directly subject to Carthage and the immediate neighbourhood of the Phœnician and Punic colonies dotted along the coast. Not until the Roman period would North Africa become one of the granaries of the Mediterranean world. None the less, Carthage inspired the first progress made by the Berbers in agriculture and stock-breeding.

The most abundant cereals continued to be, as in prehistoric times, wheat and barley; we cannot say what varieties predominated. The Phœnicians introduced fruit-growing into the country. The soil and climate were favourable to the vine, the olive, the fig, the almond, and the pomegranate, and it is possible that all these had long grown there in a wild state. Probably the arts of pruning and grafting, the special treatment required by each species, and methodical planting in ground suitably situated and prepared were now introduced. The earliest African orchards date from the time of Carthaginian rule. The exploitation of the date-palm can only have interested Carthage in the few oases scattered along the coast of the Syrtes. Round the cities, the vegetables needed to feed town populations were grown in large quantities.

Stock-raising, too, was greatly improved. The Carthaginians possessed a cavalry which, if not very big, was at least very well looked after. The grass-lands over which the army of Agathocles went in 310 were full of horses. Nor was there any lack of mules. Most of the races of domesticable animals known to the ancients were represented in Libya.

> "In that country," Polybios says, "there is such abundance of horses, oxen, sheep, and goats, that I do not think anything like it could be found in all the rest of the world."[1]

[1] Quoted in **LXXI**, vol. iv, p. 40.

One of the Carthaginian sacrifice-tariffs now known enumerates, as victims due to the deity, calves, rams, sheep, he-goats, lambs, and kids.[1] Poultry-keeping and bee-keeping were equally prosperous. Punic wax was especially renowned.

This agricultural and pastoral wealth was due, not only to the nature of the soil and climatic conditions, but to a serious study of the subject. In this respect, the economic practice of Carthage was at an early date, if not exactly scientific, at least reasoned and thoughtful. The Greeks and Romans knew and quoted authors who had written of agriculture and stock-breeding in the Punic language. The most famous of these writers was Mago, whose treatise in twenty-eight books was translated into Greek and Latin. Varro, Columella, and Pliny the Elder often refer to it. Mago did not merely deal with the treatment of plants, trees, and livestock; he touched upon every matter covered by farming. If we possessed his work, we should find valuable information about the implements and methods used in the agriculture and stock-farming of Punic Africa. What we know of it comes to very little. The plough used was simply the old implement without a fore-carriage, with a triangular iron share. The grain was separated from the straw either by being trodden by animals or by means of threshers of various types. Ingathered crops were stored in siloes. We do not know how the Carthaginians made wine, but it seems certain that they made plenty of it, and perhaps they even clarified it with gypsum. We do not know how they extracted oil from olives; no details have come down to us about the shape or working of their presses. But, although there are great gaps in our knowledge of Punic agricultural lore, it is at least certain that it was highly developed, and constituted " a true science, of which there were very learned teachers and very keen students among the aristocracy " of Carthage.[2]

To sum up, " the Carthaginians devoted themselves to agriculture with success. . . . By the exploitation of their possessions and by the influence which their example had on the natives, they contributed greatly to the foundation of the material prosperity which Africa would enjoy under Roman rule."[3]

[1] *Ibid.*, p. 41. [2] *Ibid.*, p. 8. [3] *Ibid.*, p. 2.

About the organization of rural property in the Carthaginian Empire, our information is fragmentary. Did the State keep land for itself and work it direct ? Did the estates held by the citizens of Carthage belong to them entirely, or were they burdened with taxes asserting the eminent ownership of the State ? These are questions which cannot be answered, for lack of evidence. We only know that the territories conquered by Carthage in Africa, western Sicily, and Sardinia and left in the hands of the subject peoples had to pay rent in kind and tribute in specie, both of which obligations bore witness to their legal status.

Slave labour was extensively used by the Carthaginian nobles for the cultivation of the land round their villas and the keeping of their livestock. In special circumstances, for the corn-harvest, hay-making, vintage, and olive-picking, they made use of agricultural labourers who may have been nomads from the parts of Africa which Carthage had not conquered. Small landowners and natives of the territory who had become the subjects of the victorious city tilled their land themselves. Carthaginian farming, therefore, included various methods, suited to the various forms of ownership and the natural conditions of the countries of which the Empire was composed.

Did this farming bring Carthage great wealth ? It seems to have been about sufficient for the needs of the State and individuals, but not to have furnished much for exportation.

We do not know what part hunting played in Carthaginian life. Sea-fishing was practised on the coasts of the Syrtes and near Carthage, and also in the Atlantic, south of the Pillars of Hercules. In these latter waters the fishing-boats stayed out for some time, for the tunnies caught were salted on board.

II

The Industry of Carthage

While agriculture and stock-breeding were practised, at least in a rudimentary form, in North Africa before the coming of the Phœnicians and the foundation of Carthage, industry was only established and developed by the newcomers and under their influence. One may reasonably suppose that that activity and influence were chiefly, if not wholly, confined to the urban centres.

The seeking and exploitation of the raw materials which the soil of Barbary might contain do not seem to have been pushed forward very actively so long as Carthage was independent. Building-materials were doubtless sought in the stone-quarries near the bigger towns,[1] and the forests of the country supplied the wood used in many industries. But we have no evidence strong enough to justify the supposition that the metallic resources of North Africa, the lead, copper, and iron mines now known in various parts of the country, were turned to account.[2] Like tin, these ores or metals were imported from the Peninsula or even further. Gold came from Central Africa, either by sea along the Atlantic coast or overland across the Sahara.

In every urban agglomeration, especially in a big town like Carthage, many industries which are indispensable to all collective and social life inevitably grew up, those of building, wood and stone, spinning and weaving, leather, furniture, and food. There is nothing peculiarly Punic in all this. We need only dwell on those industries which throve most at Carthage and in her colonies and shed a light on the special characteristics of Carthaginian industry.

The greatness of Carthage was based on her sea power, both commercial and military. Shipbuilding was therefore important, being conducted partly by private yards and partly by the State. Polybios tells us that the Carthaginians were very expert in this industry. They obtained wood from Africa and esparto for ropes from Spain. The improvement of harbours and the organization of special yards and workshops developed along with the progress of shipping.

The metal industries first arose in Barbary under Carthaginian rule. The Punic inscriptions mention iron-founders, copper-founders, and manufacturers of various utensils. The Punic cemeteries have yielded, in addition to some weapons, such as swords, daggers, and heads of spears, javelins, and arrows of iron or bronze, numbers of tools and implements: axes, hammers, knives, shears, scrapers, hooks,

[1] It is not certain that the quarries of Numidian marble (afterwards famous) near Simitthu (Shemtu) were opened by Carthage.

[2] **LXXI**, vol. iv, p. 49.

bells, mirrors, and little shovels, some of bronze and others of iron; copper blades of a ritual character, which some take for razors and others for small axes; bronze vessels, such as ewers, scent-pans, dishes, cups, bottles; and a few articles of lead. The metals were also used for adorning wooden chests and boxes, which were given handles and other fittings of bronze or lead.[1]

Equal use was made of the precious metals. The goldsmith and jeweller used gold and silver for decorating sanctuaries, in which we hear of crowns, tabernacles, and plates of gold; for supplying the rich with table-services, bowls, vases, and sometimes even shields, of gold or more often of silver;[2] or for satisfying the love of ornament, which was very widespread among the Carthaginians, with a profusion of rings, bangles, pendants, earrings, necklaces composed of all sorts of materials, drops, head-bands, amulet-cases, etc.[3]

To the same love of display a number of artistic industries owed their importance—those of fine stones, particularly scarabs for use as seals, of articles of enamelled paste, of glass vases, of combs, mirror-handles, fan-handles, furniture-fittings, and many small objects of ivory, and of painted or engraved ostrich-eggs.[4]

The ceramic industry was as prosperous as those of metal and the precious substances. The activity which it showed is attested not only by the innumerable pots dug up in the cemeteries but by the presence of workshops and kilns at Carthage and elsewhere, and by the marks, which are certainly Punic, stamped on vase-handles. Far the most of the articles produced by this industry were common utensils—jars, amphoras, ewers, urns, jugs, goblets, dishes, bowls, and lamps with one or two spouts, all made on the wheel and fired in the kiln, usually of a heavy, commonplace shape and painted with red, black, or brown networks, zigzags, or circles; some pieces are adorned with rather rudimentary palmettes, branches, or petals. From all this mass of uninteresting objects a few more exciting specimens emerge—pots shaped like animals and ewers with human heads; statuettes of women in a rigid attitude, figures making the gesture of prayer, and a few deities; medallions adorned with

[1] **LXXI**, vol. iv, pp. 74 *ff*.
[2] *Ibid.*, pp. 83 *ff*.
[3] *Ibid.*, pp. 85 *ff*.
[4] *Ibid.*, pp. 93 *ff*.

vegetable or animal motives; grotesque masks; and female busts.[1]

As a whole, the industrial output of Carthage is not distinguished by a single original feature in the methods of manufacture or in the shape or decoration of the articles. So far from showing any inventive spirit, it reveals a real inability to progress or to take on new life; it is affected first by the influence of Egypt and then by that of Greek Sicily. Active as they were, Punic metal-working and pottery did not contribute anything at all to the economic progress of the ancient world; they merely assisted the spread and development in North Africa of those two industries as practised for various lengths of time in the other parts of the Mediterranean.

We have very little information about the organization of industry at Carthage. We may take it that shipbuilding, at least in the case of warships, was a State industry. Carthage does not seem to have had huge industries or big workshops chiefly employing slave labour. It is probable that most trades—metal-working, goldsmith's work, jewellery, pottery—were practised either in medium-sized workshops, in which a few workmen served an employer, or by free craftsmen who might, in some cases, have one or two friends to help them.[2]

These manufactures chiefly supplied the home trade of the Carthaginian state; they also furnished articles of exchange to the shipowners and merchants who traded with the African tribes. Moreover, if some of these goods were exported abroad, Carthage imported similar articles of much better workmanship and finer style, which her ships brought from Greek lands.

III

The Trade of Carthage

We have seen that the first real progress made in North Africa by agriculture, stock-breeding, and industry was due to the Phœnician colonization and the influence of Carthage. But by far the most characteristic activity of Carthage was

[1] *Ibid.*, pp. 57 *ff.*

[2] *Ibid.*, pp. 53 *ff.*

the trade, chiefly sea-borne, which for several centuries gave her a real hegemony all over the West.

We need not dwell on the small trade which went on in the towns between merchants and local consumers. This is a practice common to all civilizations, and, besides, we have no definite information about it in the case of Carthage and her African colonies. Much more important and characteristic was the trade which the great Punic city carried on with foreign countries, near and far.

Carthage imported, chiefly from the Greek countries and the East, a number of foodstuffs and many manufactured goods for the use of her people or her subjects. Sicily (especially Acragas), Campania, and the island of Rhodes, sent her wine, and much oil came from Acragas.[1] From the Greek world, either Greece or Campania, she received, perhaps through Sicily, bronze articles, jewellery, and painted vases, among which Corinthian wares and pots made in Campania and Apulia have been found. Terra-cotta statuettes and bronze ewers from Cyprus have been discovered in the Punic cemeteries. Egypt also supplied Carthage with various articles of adornment.[2] The African territory and colonial empire of Carthage supplied very little to her export trade—slaves and ores and metals, especially the lead and silver of southern Spain.

Carthaginian trade owed its prosperity and special character to shipping much more than to importation and exportation in the strict sense of the two words.

Either on her own account or through her colonies, Carthage received on her market or went long distances to obtain goods which she then distributed in the countries with which she had commercial dealings. The trading-stations founded by the Phœnicians on the coasts of the Syrtes—Leptis Magna, Œa, Sabrata, Gigthis, Tacape—which had become allies or vassals of Carthage, were connected by caravan with the great oases of the Fezzan and Ghadames, and by those oases with the Sudan and Central Africa. Although the Carthaginians have given no definite information about the trade that went on by these routes, one is justified in supposing that from those still mysterious regions they obtained black slaves, ivory, skins of wild beasts,

[1] **LXXI**, vol. iv, pp. 27, 29. [2] *Ibid.*, pp. 154 *ff*.

ostrich-feathers, and gold. Another station to which they went for many products of Central Africa was on the Atlantic coast, beyond the Pillars of Hercules. From there they brought back skins, lions, panthers, antelopes, ivory, and gold, which they obtained in exchange for perfumes, pottery, and glass rubbish.[1] On the European coast of the Atlantic, in Galicia and Asturias, in Brittany, and in Cornwall, the Carthaginian ships loaded up with tin and lead. We do not know the details of this transit trade; we have very little knowledge of what goods were given in exchange for these materials. We can only say that, in the Western Mediterranean and on the Atlantic coasts of Europe and North Africa, the Carthaginians, following the Phœnicians, played a part in sea-borne trade like that of the Dutch in the seventeenth century.

It is easier to determine what were the routes taken by this traffic, and what were its chief centres and geographical domain. In Africa itself, those of the old populations which remained independent do not seem to have given the Carthaginian traders much custom. The Numidian chiefs perhaps bought luxuries from them, arms, jewels, etc., but the needs of these tribes, which were partly nomadic and chiefly pastoral, were very limited. The Carthaginians did not consider it necessary to open roads far into the back-country. Sales, purchases, and exchanges usually took place in the towns on the coast, which were almost all entrepôts and markets. The tracks over the Sahara which were taken by the caravans from Central Africa were not under the control of Carthage.

The true domain of Punic commerce was the sea. Merchants went to Phœnicia, Egypt, and Greece, but the chief activity of the shipowners was in the West. The documents which we possess and the archæological observations made in various regions bear witness to continuous trade relations between Carthage on the one side and Greek Sicily, parts of Italy, such as Campania, Latium and Etruria, and certain points on the coast of south Gaul and north-east Spain on the other. Here the Greeks and perhaps the Etruscans were serious rivals. Beyond the Pillars of Hercules, the Punic ships almost had the sea to themselves. The voyages

[1] *Ibid.*, pp. 141 *ff.*

of Himilco and Hanno, which were official expeditions organized by the State, leave no doubt as to the interest which the Carthaginian government took in those distant waters. Himilco was instructed to explore the European coast of the Atlantic north of Gades, and to take steps that the carriage of the ores and metals abounding on those coasts should be as far as possible kept in Carthaginian hands. Hanno, after passing through the Pillars of Hercules, sailed southwards along the African coast, and modern historians admit that he reached the head of the Gulf of Guinea and came near to the Equator. He founded several colonies on the coast of Morocco, the southernmost of which was planted on the island of Cerne, between Capes Juby and Bojador.[1]

Carthage also took energetic steps to keep the monopoly of trade in the Atlantic and all along the Mediterranean coast of Africa. In the Tyrrhenian Sea, in the Gulfs of Genoa and Lions, and along eastern Spain, she could not evict the Greeks, but she succeeded in keeping them out of all regions over which her political authority or commercial supremacy prevailed. At the beginning of the Punic Wars her commercial domain in the Mediterranean and Atlantic was very large.

To exploit that domain, Carthage had a material equipment, some elements of which are fairly well known to us. The merchant navy was renowned for the size of its vessels, large galleys driven by sail and, if necessary, by oars, and for the skill of its crews and commanders, who were not content to hug the shore but took to the high seas, observing the stars.[2] This fleet found well-chosen and well-fitted points of call, shelters, and bases in numerous harbours. First, there was the merchant harbour of Carthage; its exact position and arrangement have given rise to many controversies, but it was certainly very large and very active.[3] Then there were the harbours of the many colonies founded by the Phœnicians and Carthaginians, which stretched along the north coast of Africa and the south of Spain, wisely placed on a strait between the mainland and an island, or under the

[1] On the expeditions of Himilco and Hanno, see **LXXI**, vol. i, pp. 468 *ff*.

[2] **LXXI**, vol. iv, p. 111.

[3] On the port of Carthage, see **LXXI**, vol. ii, pp. 38 *ff*.

shelter of a headland, or at the mouth of a river. The most important of these seem to have been Leptis Magna (Lebda), Tacape or Tacapas (Gabes), Thapsus (Ras Dimas), Hadrumetum (Susa), Utica, Hippo Diarrhytos (Bizerta), Hippo Regius (Bona), Saldæ (Bougie), Iol, afterwards Iol Cæsareia (Shershell), Gunugu (near Guraya), Tingi (Tangiers), and Lixus (near Larash) in Africa, and Gades (Cadiz) in Spain.

These ports and many other stations situated inside or outside the limits of the Carthaginian Empire served Punic commerce as entrepôts and trading-stations. Sometimes business took the form of the exchange of goods, or barter. Herodotos and the *Periplus* of Scylax say so definitely of the west coast of Africa.[1] It was the same in the ports of the Syrtes where the caravan-routes of the Sahara ended and in the stations on the European coast where the Carthaginians obtained tin and lead.[2] The use of money in commercial transactions seems to have been fairly late. Yet Carthage did not lack gold or silver; she obtained the former from Central Africa and easily procured the latter in Spain. It is possible that in their dealings with civilized peoples the Punic traders for a long time used ingots, which were weighed, or foreign coins, chiefly Greek.

It is agreed that Carthage struck no money of her own before the fourth century, or at the end of the fifth at the earliest. The first Punic coins were minted in Sicily, on the model of the Greek coins of that country. At Carthage itself and in Africa, minting did not start before the middle of the fourth century, and the coinage was inferior in quality to that of Sicily. There were Punic mints in Sardinia for bronze and in Spain for silver. It is curious that no Carthaginian coins should have been discovered outside the territories politically subject to Carthage; those found in eastern Algeria did not, perhaps, circulate there before the Roman period.[3]

In spite of defects in monetary organization, movable wealth, chiefly acquired by trade, was very considerable at Carthage. Business was held in high esteem. The aristocracy considered it no disgrace to devote its resources and energies to trade. Many nobles were shipowners or

[1] **LXXI**, vol. iv, pp. 141 *ff*. [2] *Ibid.*, p. 130.
[3] **LXXI**, vol. ii, pp. 324 *ff*.; vol. iv, pp. 130, 135.

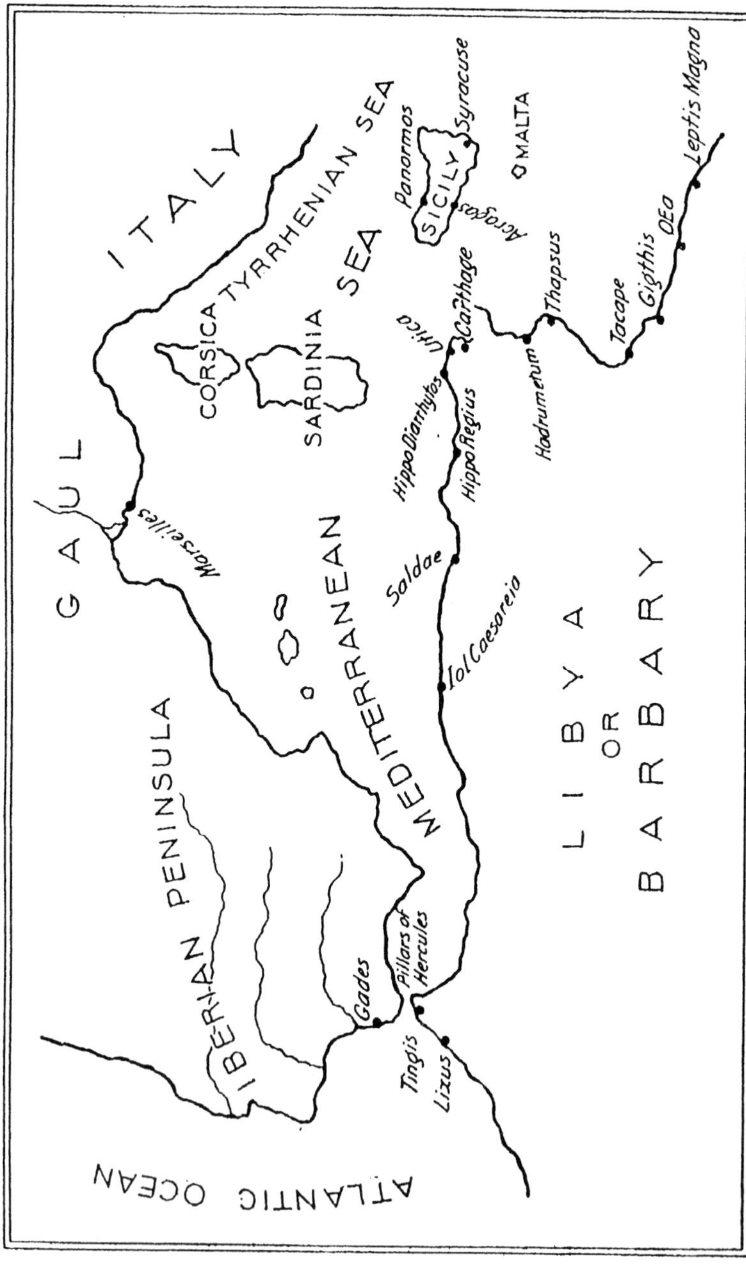

MAP IV.—THE CARTHAGINIAN EMPIRE.

bankers. Carthage was one of the ancient cities in which capitalism was most powerful and weighed heaviest on the destiny of the nation. Hannibal seems to have seen this after the defeat of Zama, for he strove, by energetic measures, to deliver the State from the tyranny of the financial magnates. But it was too late. Fiscal organization had been in their hands too long. This very fact made the loss of Sicily, Sardinia, and Spain irreparable.

What proves, moreover, the supreme place which trade held in the private and public life of the Carthaginians is the general trend of the education of the young and the whole policy of the city.

The qualities which they endeavoured to develop in the young were, first and foremost, a love of undertakings in distant lands, smartness, if not roguery, in business, eagerness to make money and a passionate love of wealth, versatility combined with tenacity, a political intelligence capable of profiting even by apparently unfavourable circumstances, obsequiousness to the powerful, and merciless arrogance towards the weak. The Latin writers may perhaps have exaggerated the shortcomings of Punic honesty, *fides Punica*, but they did not invent them. The Carthaginians inspired dislike among the peoples of antiquity as a whole; the Greeks speak of them just as severely as the Romans.[1]

The general policy of Carthage seems to have been mainly inspired by commercial interests. "The Republic," M. Gsell writes, "had a commercial policy, which may be summed up as follows: to open up markets for the Carthaginians by force or by treaties or by the foundation of colonies; to keep these markets to themselves in countries where it was possible to ward off all competition; where this was not possible, to govern transactions by agreements specifying reciprocal advantages; and to protect the freedom of navigation and the existence of cities and seaside trading-stations against pirates."[2] Precious evidence of that policy survives in the two commercial treaties concluded between Carthage and Rome. The dates of these agreements may be disputed, but their existence and their purport are indisputable. The text has been preserved by Polybios. According to M. Gsell, the earlier of these treaties goes back to the end of the sixth

[1] **LXXI**, vol. iv, pp. 217 *ff*.

[2] *Ibid*., p. 113.

century, while the second was signed about the middle of the fourth.[1] While the Carthaginians undertook to respect the coast of Latium and central Italy, not to attack any city there and not to build any stronghold there, they forbade the Romans to do any trade, or almost any trade, in the regions under their control, along the African coast from the Syrtes to the Atlantic, on the south coast of the Iberian peninsula, or in Sardinia. In the second treaty we find the following clauses:

"In Sardinia and in Libya, no Roman shall do trade, or found cities, (or land) longer than is needed to take on provisions and repair his ship. If he is driven there by a storm, he shall depart within five days. . . . Beyond the Fair Promontory[2] and Mastia in Tarseion[3] the Romans may neither seize plunder nor do trade and may not found cities."[4]

It is likely that similar agreements were made by the Carthaginians with the Greeks, and perhaps also with the Etruscans. "From the sixth century onwards," M. Gsell concludes, "the Carthaginians established commercial monopolies in the West. In the fourth century, they allowed no competition in Africa west of Cyrenaica, in Sardinia, in the south of Spain, or beyond the Strait of Gibraltar."[5] Only the port of Carthage was open to foreign trade, with serious safeguards, but under a real control; the same conditions were granted to the Romans in the ports owned by Carthage in the west of Sicily.

Piracy was practised by all sea-faring peoples. The danger was especially serious for Carthage, whose wealth and power were founded on marine trade. The Punic State put it down severely. Various clauses in the treaties concluded between Rome and Carthage suggest that the two cities mutually undertook to abstain from such aggression.[6] Similar undertakings were made between the Carthaginians and Etruscans.

Although there are many gaps in our knowledge of Punic commerce, it is at least certain that trade, chiefly sea-borne, was by far the most important element in the economic

[1] **LXXI**, vol. i, p. 461.
[2] Cape Sidi Ali el-Mekki, N. of Carthage.
[3] In Spain, near C. de Palos and the site of the present Cartagena.
[4] **LXXI**, vol. iv, pp. 118 *ff.*
[5] *Ibid.*, p. 122.
[6] *Ibid.*, pp. 125 *ff.*

activity of the Carthaginians. To business Carthage owed her prosperity; through business she played a big part in the history of the Western Mediterranean; business gave her her special character among the great cities of antiquity. It would be absurd to deny the influence which, by her commercial relations, she was able to exert on the destinies of North Africa. But her history shows the superficiality and weakness of a civilization in which the chief driving-power of human activity is the conquest of riches, and there is no endeavour to achieve, by the side of and by means of economic power, political, intellectual, and moral progress.

CHAPTER III

THE ECONOMIC LIFE OF PRIMITIVE ITALY. THE ETRUSCANS. ROME TO THE MIDDLE OF THE THIRD CENTURY BEFORE CHRIST[1]

THE economic evolution of Italy, in the earliest period of its history, is more complex than that of Western Europe and North Africa. No doubt, one finds the characteristic ages of prehistory in their accustomed order, Palæolithic, Neolithic, Copper, Bronze, Iron. But in the course of these great stages in progress, towards the rational use of natural resources, towards the methodical and fruitful organization of human labour, Italy underwent many migrations of various origins and was subjected to influences as varied as they were numerous. By its general shape, stretched out from north to south from the Alps to the Strait of Messina, by its position between the two great basins of the Mediterranean, by its three fronts on the Adriatic, the Ionian Sea, and the Tyrrhenian Sea, and by its very relief and its topography, which favours division rather than unity, Italy belongs both to Central Europe and to the Mediterranean world. It faces both east and west; the peoples which occupied it and the civilizations brought to it from outside or developed on its own soil did not merge into a truly national unity until very late. There are, moreover, obscurities and gaps in the prehistory of Italy. To dispel the former and bridge over the latter hypotheses have been suggested and systems erected which do not always appear very solidly founded.

Here, at least, are the main features of the process. After the Palæolithic period, many traces of which have been found from the Po valley to Sicily, the Neolithic Age seems to have been of great importance in Italy. It was the inhabitants of

[1] For the principal parts of this chapter I have chiefly made use of A. Piganiol, *Essai sur les origines de Rome*, Paris, 1917, and L. Homo, *Primitive Italy*, London and N.Y., 1927, in which very many works of specialists are mentioned and used; *cf.* **LXXII**, pp. 64 *ff.*; **LXXVII**, *passim*; **LXXIX**, pp. 31 *ff.*, 241 *ff.*; **LXXX**, *passim*.

that period whom ancient tradition designated by the names of Ligurians and Sicels and regarded as the most ancient people of Italy. These populations were of different races, and the word "Ligurian" must be used to designate a state of civilization rather than a racial element.

At the end of the Neolithic Age, the metals appear, first copper and then bronze. North of the Apennines, numerous palafittes or lake villages are built in Lombardy and Venetia, and terramare, villages on piles built on dry land and surrounded by a moat full of running water, appear, especially in Emilia, from Piacenza to the neighbourhood of Bologna. Most prehistorians hold that the inhabitants of the lake villages and terramare came to Italy from Central Europe and formed the advance-guard of the Italici properly so called. The theories according to which they were Ligurians, or Etruscans, or even Celts, are now abandoned, or very much contested. The Bronze Age with its lake villages and terramare was followed by the Iron Age, represented in Italy by the Villanovan civilization. This was brought in by the second line of the great Italic migration, the Umbrians and Sabellians, by whom the immigrants who had previously come down from the Alps to the banks of the Po and its tributaries were driven onwards into Central Italy. The Celts did not appear in that region till much later.

While numerous populations were entering Italy from the north, other bands were coming in from the east, from the Balkan Peninsula, starting from Illyria. Some of these bands went round the head of the Adriatic by land, others crossed that sea at the middle and landed on the coast of Picenum, and others seem to have come by the Strait of Otranto. The Veneti in the north, the Picentines and Peligni in the centre, and the Iapygians in the south are generally regarded as being of Illyrian origin.

These various migrations, from Central Europe by the passes of the Alps and from Illyria by land and sea, had come to an end about the beginning of the first millennium before Christ. Already the civilization of Italy had been affected by Eastern and Mediterranean influences. In the Neolithic Age we find the influence of pre-Minoan Crete in Sicily, and it grows stronger during the Copper Age. In Villanovan times, north-eastern Italy received from still further, from Cyprus and Phœnicia as well as from Greece, if not models, at least inspirations which affected local industries.

All this was only the very modest prelude to a much more extensive and much profounder action. Two peoples, two civilizations from the Eastern Mediterranean, arrived in Italy between the end of the eleventh century and the eighth—the Etruscans and the Greeks. In the days of its greatest expansion, the Etruscan power, whose centre always remained in Tuscany, extended from Campania to the valley of the Po. The field of action of the Greeks, even if we admit that they founded colonies, which afterwards disappeared, in central and northern Italy, on the Etruscan coast, and at the head of the Adriatic, was chiefly southern Italy and Sicily. But Great Greece and Sicily were for hundreds of years an integral part of the Hellenic world, and their economic history belongs in the main to that of Greece.[1] Their share in the economic life of Italy was chiefly marked by the trade relations which they maintained with Etruria and Latium and the influence which they had on the civilization of those two regions. The economic history of the Etruscans, on the other hand, belongs to the history of Italy, and it is not too much to say that before Rome arose Etruria gave the economic life of Italy its first effective and native impetus.

The primitive period of Italian economic life therefore ends with the arrival of the Etruscans in the peninsula. What were the general characteristics of that life before this capital event took place ?

I

The Economic Life of Primitive Italy

Primitive Italy passed through a succession of economic phases very similar to those which we have seen in Western Europe. Palæolithic man lived by hunting, fishing, and gathering wild fruit. His industries were confined to chipping stone and preparing bone, horn, and perhaps wood, skins, and leather. The first real progress was made in the long Neolithic Age. It is hard to say whether agriculture was practised in Italy from the very beginning of this

[1] See above, pp. 24 *ff*.

period,[1] but at least one of man's greatest conquests over hostile nature then took place—stock-breeding appears. In any case, when the Neolithic Age ended, in the Copper Age, the chief forms of economic activity already existed. Man had domesticated several races of animals and had learned to till the soil; he practised several industries previously unknown, including pottery and metal-working; and he had come into relations with Crete and the Ægean civilization, from which he doubtless received manufactured goods.[2]

This progress became more marked in the time of the lake villages and terramare. Here we can come down to details, thanks to the abundant evidence found on the sites of these primitive dwelling-places. "The inhabitants of the Terramare, a numerous and flourishing people judging by the remains they have left, mark a stage of culture much higher than that of the Neolithic Age. While hunters and fishers like their predecessors in Italy, they were primarily an agricultural people. . . . They cultivated edible and useful plants (wheat, beans, vines, fruit-trees, and flax) and pastured their flocks and herds (cattle, swine, sheep, barnyard animals). They carried on a textile industry (weaving for clothes and the plaiting of twine), wood-working (axe-hafts, baskets of withies, spades, knives, chisels, polishers, dishes, basins, ladles, straight or curved staves, and bows), bone-working and horn-working (needles, knives, hammers, chisels, combs, hair-pins, and spindle-whorls), and a ceramic industry (pottery, unbaked or baked at the open fire, and distinguished especially by a crescent-shaped projection on the handle and its decoration, lines, furrows, or 'kicks'). Finally, a bronze industry characterized this new civilization. Bronze was used for arms (axes, daggers, knives, arrow-heads, and swords) and implements (razors, sickles, blades, tweezers, hair-pins, and combs), and, finally, for ornaments (little wheels, pendants, and violin-bow fibulæ). The metal was cast in a mould, but technique was still primitive."[3] The same civiliza-

[1] M. Piganiol seems to say that it was: "The Sicels were a people of agriculturalists. . . . The Ligurians were a race of peasants who tilled the ground and buried their dead" (**LXXXI**, pp. 11-12). M. Homo, on the contrary, makes agriculture first begin with the Copper Age (**LXXIII**, English p. 29).

[2] **LXXIII**, English pp. 26-29.

[3] *Ibid.*, English p. 32.

tion developed in the Italian palafittes of the Venetian lakes, in Tuscany, in Campania, and in Sicily.[1]

In Italy as elsewhere in Europe, the Bronze Age was followed by the Iron Age. The best known, most abundant, and most fully studied evidence of that new form of civilization is the deposit of Villanova, in the Bologna district.[2] Economically, Villanova is chiefly distinguished from the terramare by the progressive introduction of iron into daily life; little by little the new metal took the place of bronze for the manufacture of weapons, tools, and articles of luxury and adornment.[3] It is possible that the working of the iron-ores of the Tuscan coast and the island of Elba dates from this period.

The civilization brought into parts of Italy—Venetia, Picenum, Apulia—by the Illyrian invaders is not earlier than the Iron Age. It differs in some respects from that of Villanova, but the economic practices of the two are similar. M. Piganiol notes two interesting details in respect of Picenum—the abundance of iron and scarcity of bronze, and the huge abundance of amber.[4]

So, before the Etruscans and Greeks came to Italy, the economic development of the country had been mainly determined by influences coming either from Central Europe across the Alps or from the Balkan Peninsula. No migration had so far brought truly Mediterranean or Eastern racial elements into the peninsula. It seems, however, that even in those early times there were commercial relations between Italy and the pre-Hellenic East. In the Neolithic Age, Sicily already knew the products of pre-Minoan Crete.[5] During the Copper Age, the influence of the Ægean civilization became more marked. At the same time the people of the lake villages and terramare, being bronze-users, must have been in relations with the countries which produced tin or with the stations dotted along the road by which that metal came to the Mediterranean. Later, part at least of the iron used by the Villanovans and their contemporaries came without any doubt from the mines of Central Europe. The presence of many amber objects in the Picentine cemetery

[1] **LXXIII**, English pp. 32-33.
[2] **LXX**, *passim*.
[3] **LXXIII**, English pp. 33 *ff*.
[4] **LXXXI**, pp. 24 *ff*.
[5] **LXXIII**, English p. 27.

of Novilara bears witness to the importance assumed in Italy by the trade in this commodity.

Whatever may have been the subsequent importance of the Etruscan immigration and the Greek colonization on the destinies of Italy, the country in which these two peoples landed was no longer, at the beginning of the first millennium before Christ, a primitive country economically. Agriculture, stock-breeding, industry, and trade had long been in existence there; human activity was exercised in a variety of fruitful occupations. The ground was well prepared to receive the seeds brought by the new-comers from the East and Greece.

II

The Etruscans and their Economic Life

In spite of the discoveries of archæology and the studies of many scholars, the history and civilization of the Etruscans are still in many respects obscure and enigmatic. Some important facts are, however, regarded as beyond dispute. It is almost universally agreed today that the Etruscans came to Italy by sea, that they came from the Eastern Mediterranean, and more exactly from the coast of Asia and some islands in the northern Ægean, and that they landed on the west coast of the peninsula between the mouths of the Tiber and the Arno. The migration took place about the tenth century B.C. First they occupied the country lying between the coast of the Tyrrhenian Sea and the Apennines, Etruria properly so called, now Tuscany; then they extended their power southward to the northern borders of Campania, and north-eastward beyond the Apennines all over the middle and lower valley of the Po. Between the seventh and fifth centuries, the Etruscan Empire stretched over northern and central Italy from the foot of the Alps to the Silarus. Melpum (the future Mediolanum), Placentia, Mantua, Verona, Ravenna, and Felsina (afterwards Bononia) in the north, the twelve chief cities gathered round Vulsinii and afterwards Rome in the centre, and Cumæ, Capua, Naples, and Pompeii in the south all came under the sway and almost all under the influence of Etruria. Not until the fifth century did its power decline, when the reaction of Latium and

Rome, supported by the Greeks of Cumæ, gradually robbed it of all the land which it had occupied south of the Tiber, and the Gallic invasion of northern and north-eastern Italy drove it back across the Apennines. In the fourth century, "Etruria was in full decay, and ceased to count seriously in Italy."[1]

But its influence on the general civilization and the economic life of ancient Italy was intense and profound. The regions over which it ruled were truly transformed by it, and the progress which it set going was not undone nor even injured.

The Etruscan civilization was essentially of an urban character. The terramare and the mere agglomerations of dwellings of the Villanovan period were now succeeded by true cities, surrounded by strong walls, built on a strict geometrical plan, and often crowned by an acropolis. In these cities buildings of stone, baked brick, or *pisé* (beaten earth) made their appearance, whereas the dwellings of the previous periods had been made of branches, straw, and mud. Like the abodes of the living, the tombs of the dead were properly built chambers, varying in size, with walls of massive stone blocks and solid pillars to support the roof.

The Etruscans were not only architects; they were also, and perhaps mainly, remarkable engineers. Ancient tradition regarded them as the authors or inspirers of the drainage-works which for hundreds of years made the Tuscan Maremma, the Roman Campagna, and the Pontine Marshes relatively healthy spots. It was to the time called the age of the Tarquins, that is, the period during which Rome was ruled by the Etruscans, that men ascribed the building of the sewer, or rather drainage-channel, known as the Cloaca Maxima, by which the swampy valleys round the Palatine were drained. Outside the towns, the Etruscans cleared forests, reclaimed much land from the stagnant water which covered it, dug quantities of little channels or *cuniculi* which carried off water which lay on or immediately under the surface of the ground, and, by a hydraulic feat which may serve as a model at this day, brought into use or taught the native populations to bring into use, without any serious risk to their health, a soil which by nature was hostile and

[1] **LXXIII**, English pp. 159-60.

unhealthy.[1] Elsewhere, for example along the Po and its chief tributaries, they built embankments and cut ditches to hold, direct, or slow down the flow of water coming from the Alps or the Apennines.[2]

Agriculture then developed marvellously in the countries subject to Etruscan influence. We have no details enabling us to describe the various methods and the results of that activity, but can at least judge of its importance when we note that Rome afterwards exacted heavy contributions in corn from the Latin and Etruscan cities which she conquered.

Industry also made some progress, but the Etruscans seem to have introduced fewer novelties in this domain. They opened up many large quarries from which they obtained the materials needed for building their city walls, public buildings, and dwelling-houses. They doubtless gave a great impetus to the extraction of ore on the Tuscan seaboard and in Elba. The presence of those ores is still attested by the modern names of Monte Argentaro, Piombino, and Porto Ferraio, and they may have been known as early as Villanovan times. From the forests of Corsica, after the departure of the Phocæans, and those which covered the Apennines they obtained plenty of timber. The various building industries, principally masonry and heavy carpentry, throve in Etruria. Pottery developed especially. It is true that most of the vases found in the cemeteries are of Greek origin, and do not therefore represent the national industry of the Etruscans; the use of terra-cotta in the decoration of buildings, the modelling of full-sized statues of terra-cotta, and the manufacture of polished black vases (bucchero) all seem to have been Etruscan inventions. They certainly did not introduce pottery into Italian industry, but, thanks to their encouragement, terra-cotta won a far greater place in it than before.

It was the same with metal-working. Among the many bronze objects of which the funerary furniture laid in the tombs was composed, some no doubt were bought by the Etruscans in Greece; but they also made them, with less

[1] R. de la Blanchère, *Un Chapitre d'histoire pontine*, *passim*. The author of this remarkable work nowhere mentions the Etruscans; but there is good ground for believing that the "cunicular drainage" described in such detail was done or inspired by them; *cf.* **XVII**, *s.v.* "Cuniculus."

[2] **LXXIII**, English p. 110.

art and delicacy, of course, but with undeniable technical skill. All this bronze-ware, cast, repoussé, or adorned with incised patterns, points to the progress made by metallurgy in Etruria.[1]

Bone and ivory were also worked.

Things made by the Etruscans themselves have not only been found in Etruria properly so called; they have been discovered in Rome and Latium on one side and in the Po valley on the other. This diffusion enables one to estimate the importance of the influence exerted by the Etruscans on the industry of ancient Italy.

The influence of their commercial activity was considerable. By the mere extension of their empire, by the extent of territory subjected to their sway, the Etruscans were led to maintain an active trade with the Greek colonies of southern Italy and with the peoples of the Alpine regions, both in the southern valleys and on the northern slopes of that range. But most of their trade was done by sea. Although conditions on the Tuscan coast were not favourable to navigation, and although even those of their cities which were nearest to the sea, Cære, Tarquinii, Vulci, Vetulonia, were not actually on the coast, they obtained a naval power in the Western Mediterranean which enabled them to hold their own against the Greeks and to treat Carthage as an equal. On the Adriatic north and south of the delta of the Po, Hadria, Spina, and Ravenna were busy ports.

The discoveries made in the Etruscan cemeteries, both north and south of the Apennines, have proved that the commercial dealings of Etruria were chiefly with the Greek world. The importation of painted vases and bronze articles seems to have been the most important part of this trade. The potters and bronze-workers of Corinth and Athens were the chief suppliers of Etruria. It is not known whether Etruscan ships went to get these precious wares in the producing countries; perhaps the Greeks of Sicily brought them. Nor have we exact details about the nature and amount of the trade which Etruria did with Carthage. We

[1] One centre of the Etruscan metal industry seems to have been Arretium. In 205, when Scipio was starting for Africa, Arretium provided him with 3,000 shields, 3,000 helmets, 50,000 darts, javelins, and long pikes, besides axes, picks, scythes, etc. (Livy, xxviii, 45).

do not know how its commerce was organized. We do not know what goods the Etruscans transported to their neighbours, nor to what extent they used money. Etruscan epigraphy and numismatics are still very obscure.

Much as we may regret the gaps in our knowledge of the economic life of the Etruscans, we know enough of it to be able to appreciate the great part which it played in the economic history of Italy before the country was united under Roman rule.

Whatever the relations of the centre of the peninsula and the Po valley may have been with the Ægean and Eastern world in prehistoric times, it was chiefly thanks to the Etruscans that those regions became acquainted with Greek civilization and first felt its beneficent influence. Not only did they bring into the country agricultural and industrial methods previously unknown, which represented an enormous advance upon the habits and processes of earlier ages, but, by constantly keeping in contact with Hellenism in the most brilliant centuries of its history, they caused Greek influences to spread more and more among the peoples of Italy—influences which were as potent in economic matters as in intellectual and moral matters. It was under the guidance of Etruria that Rome herself took her first steps on that road. When the Etruscan power declined, when the Etruscan Empire collapsed, and Rome gradually became the most powerful city in Italy, she owed her triumph in great part to the fact that she had received the seeds of her economic future from Etruria.

III

The Economic Life of Early Rome

The city which was to preside for several hundreds of years over the economic synthesis and expansion of the ancient world did not begin to rise to that magnificent destiny until very late. The oldest inhabitants of Latium, in particular those who occupied the swamp-girt hills on which Rome would one day rise, moved more slowly along the road of economic progress than the peoples established in the valley of the Po, to say nothing of the Etruscans. The so-called Latian civilization, which was contemporaneous and related with that of the terramare, was very little affected by Villanovan influence. Down to the time when, thanks to the neighbourhood of Great Greece and the Etruscan domination, Latium began to feel the beneficent effects of contact with two higher forms of civilization and economic activity, the earliest Latins, or Prisco-Latins, were on the whole "still only poor, half-nomad shepherds, living a rude life and still observing often ferocious customs, if we may judge from the typical survival of the *rex nemorensis*, 'the king of Nemi.' They were acquainted neither with writing nor, probably, with the agriculture in which they were to be instructed by the Etruscans. That is the . . . environment in which Rome was born and grew up for centuries."[1]

The history of the beginnings of Rome has gradually been brought to light by modern criticism and the archæological discoveries made within the last twenty-five years on the site of the city. There still remain many uncertain details and belts of darkness. It is, however, possible to perceive the chief stages of her development and to obtain exact economic data from the facts known. At the beginning of the history of Rome, the Aventine was the seat of a human group, composed chiefly of hunters and herdsmen, settled in a very humble way among the woods and in the few clearings of the hill. Somewhat later the village of Cermalus or Germal was built, on the west end of the Palatine. Then the neighbouring hill-tops and brows were occupied, in what order, one cannot say. The eastern part of the Palatine was covered by another village, Palatual; four were established on the Esquiline and Cælian; the Velia, a ridge connecting the Esquiline with the Palatine, was inhabited. These modest communities drew together; so the Septimontium was formed, the League of the Seven Hills, the organization of which appears as an intermediate stage between the primitive, isolated village of Cermalus and the city of Etruscan and Servian times, which is often called the City of the Four Tribes. The Aventine on the one side and the Capitol, Quirinal, and Viminal on the other remained outside the League of the Seven Hills; they were not, however, uninhabited, for they bore villages like the other hills of the future Rome.

[1] **LXXIII**, English p. 75.

Before all these villages were concentrated and combined in a single organism, a true city, which was to all appearance the material and administrative achievement of the Etruscans, their economic methods were still rudimentary. Stock-breeding, the principal occupation of the inhabitants, supplied them with their chief resources. Words like *pecunia*, obviously derived from *pecus*, and the name of Mugonia given to the gate of the Palatine opening on to the Velia bear witness to the importance of cattle in the daily life of that distant time. The ground seems to have been tilled only enough to meet the needs of the family.[1] The dwellings were mere huts, rectangular, circular, or elliptical, with walls of branches or reeds coated with badly baked clay. The industry revealed by the earliest cemeteries excavated on the site at first consisted in the manufacture of a clumsy pottery and weapons, tools, and utensils of bronze, more rarely of iron. However, one sees some progress. The use of the wheel enabled the potters to turn out vases which were less crude and to invent and execute shapes which were less barbaric, and the use of iron spread more and more. At length the economic isolation of Latium came to an end. The first relations with the Hellenic world are betrayed by the importation of vases adorned with geometrical patterns or animals, small objects such as beads, whorls, and statuettes of glass paste, and disks of amber. It was becoming clear that there were advantages in the geographical position of the spot round which were gathered the villages on the *montes* and *colles* which would one day be surrounded by the walls of Rome. There two routes of the greatest economic importance intersected—the River Tiber, which led up from the Tyrrhenian Sea into Latium and a large part of central Italy, and the land-road which connected the north of the peninsula with the south, running through Etruria, Latium, and Campania.[2] It is not impossible that even at this time there was a village at the mouth of the Tiber, where the port of Ostia afterwards arose, but it is unlikely that it was in any way dependent on those which stood some miles up the river.[3]

A new period in the general and economic history of Latium and Rome began with the Etruscan expansion.

[1] *Ibid.*, English p. 84. [2] **LXXV**, pp. 24 *ff*.
[3] **LXXIII**, English p. 96.

Careful as Roman tradition was to conceal, distort, or soften down anything in the past which might detract from the pride and glory of Rome, it admitted that the Etruscans had ruled for a fairly long time over Latium and, in particular, over the collection of villages by the Tiber which occupied the site of the future capital of Italy. It credited the Etruscan Kings with two works of capital importance in the history of the city—the Cloaca Maxima and the stone fortifications known as the Wall of Servius. Here again, modern archæology and criticism have greatly added to the information provided by the Roman annals.

It was the Etruscans who really created the *city* of Rome, both in hard bricks and mortar and in political and administrative organization. By draining off the water of the swampy ground between the various hills into the Tiber, and by enclosing in a single wall all the inhabited hills from the Quirinal and Viminal in the north to the Aventine in the south and from the Esquiline and Cælian in the east to the Capitol in the west, they founded a solidly constituted urban organism at a point which was admirably chosen not only from a military and political point of view but also from an economic one.

The mastery of the Etruscans in the matter of public works and engineering, which was always acknowledged, was doubtless also applied, although no ancient evidence expressly mentions the fact, to the reclamation and working of the land of the Roman Campagna and the neighbourhood. Then the new city, in addition to the old resources, which consisted chiefly in stock-breeding and forestry, had agricultural produce—corn and fruit, especially olives and vines. Agriculture came to enjoy a real prosperity, in a large district which was afterwards the domain of malaria.

The birth of a new city and the rise of agriculture in all the surrounding territory attracted foreigners to Rome, craftsmen and traders for the most part. This was, at least to some extent, the origin of a social organization which had serious consequences on the economic development of the Roman State. Moreover, under the encouragement of the political masters of the country, industry made great progress. The huts of wood and reeds were replaced by stone houses; public buildings arose, for example the oldest temple on the

Capitoline; the decoration of these buildings, fragments of which have been found, bears witness to the advance of the ceramic industry in Rome itself. Iron-working doubtless developed in the same period, for the manufacture of arms of offence and defence was necessary in a place of such strategic importance. In short, as soon as real town life appeared it gave birth to and fostered every form of labour which is closely connected with it.

Commercial activity developed simultaneously. In this respect, Rome, now one of the chief towns of the Etruscan Empire, benefited by its geographical position between Etruria and Campania. It stood on the great land-road connecting the two countries, and was perhaps the most important station on it. Ships going up the Tiber moored at the foot of the Aventine, on the left bank, where the ground was suitable to all the operations which make up the activity of a port.

Rome was now in being. Poor villages scattered over hills, which were separated by unhealthy and barren low ground, had given place to a healthy city, protected by a strong wall, surrounded by a territory which was methodically farmed, and inhabited not only by tillers of the land and stock-breeders but by workers and traders—ready, in short, to profit by all the favourable economic conditions which its position in the heart of central Italy gave it.

The decay of the Etruscan power, which had been the pioneer of this early progress, did not damage the future of Rome. But that future was to lie on very peculiar ways, determined by the external history of the city.

The history of Rome in the first centuries after the decline of Etruria is certainly overladen with legends in the Roman annalists, and we cannot accept all the details taken from them by Livy, Dionysios of Halicarnassos, and their successors. Modern criticism, even the mildest and most cautious, discovers in the official tradition repetitions, duplicate versions of one story, and episodes of subsequent invention, in brief, a whole body of reconstruction and interpretation, the accuracy and foundation of which are very disputable. In spite of these obstacles, it is possible to trace the rise of the Roman city through its most important stages. In her advance to leadership there were halts, and even retreats.

Rome was not always victorious. She underwent the Gallic invasion after the battle of Allia; her glory was dimmed by the disaster of the Caudine Forks; she had to pull herself together at the menace of Pyrrhos, the ally of Tarentum (Taras). But none of these set-backs was decisive, none of these dangers was fatal. Through changes of fortune which sometimes brought her near to extinction, Rome extended her sway over Latium; over the bordering districts, the countries of the Volsci, Hernici, Æqui, and Sabines, and southern Etruria; then over all central Italy from one sea to the other, the rest of Etruria, Umbria, Picenum, Campania, and Samnium; lastly, over the south from Cumæ (Cyme) and Naples to Rhegium and Tarentum. These progressive conquests, which brought more and more territory under the power of Rome and made her the mistress of the whole peninsula by the middle of the third century B.C., were bought at the cost of continual, almost annual fighting. In addition, the possession of a coast-line as extensive as that which runs round Italy from the mouth of the Arno on the Tyrrhenian Sea by the Strait of Messina and the Iapygian Promontory to Umbria gave Rome a power on the sea which supplanted that of the Etruscans and Greeks and would presently come into conflict with that of Carthage. The wars which she had to maintain in order to extend her political domain in this way and the actual results of her victories had great and far-reaching effects on her economic life.

First of all, the military organization of the city contributed to that development. The Roman army was not at that time, as it became later, a professional army; it was a militia, whose service was on principle temporary. The citizens were called up at the beginning of each campaign, and at the end they returned to their homes, without receiving any pay. They were liable to these military duties from the age of seventeen to that of forty-five. The strength of the legions lay chiefly in the class of small and middling landowners and free tenant farmers. Every man had to provide his own equipment. Let us try to see the economic consequences of this state of things. In the first days of spring, news is brought that the Volsci, the Æqui, the Sabines, or the Etruscans have attacked Roman territory. The Consuls raise one or more legions, according to the strength of the aggressors. The citizens thus recruited must at any moment leave their fields and their beasts. Since they get no pay, the cost of the campaign falls on every man. They may be killed, wounded, or taken prisoner, and if they come home it may be as victors or as vanquished. Even if they return safe and victorious, they often find their farms neglected, since no one has been able to do the work in their absence; their fields may even have been ravaged and their cattle driven off by the enemy.

This calamity prevented them, as we shall see later, from sharing in the profits of victory, when the war ended favourably at Rome. So they were without resources. But they must live till the next harvest. So they borrowed, and fell into debt. In Rome the laws on debt were very hard. Although we know no case of a debtor who, being unable to pay at the term, was put to death or sold as a slave by his creditor, it is at least certain that he fell, by the procedure of the *nexum*, into an absolute bondage, and that, being unable to pay his debt in kind or coin, he had to do so by working for his creditor on the hardest and most humiliating terms. "The *nexi*, delivered helpless to the greed and cruelty of their creditors, saw no term to their misery."[1] The details which the ancient historians give us about the problem of debt in Rome and the various solutions which were attempted are difficult to accept without criticism. But this was, without any doubt, one of the causes which contributed to ruining the middle class and swelling the mass of the poor, the landless, the proletariate.

Another cause of the same economic and social evolution was the manner in which victorious Rome treated all or part of the conquered peoples. We know that a fairly large number of towns in Latium, the names of which were known and have been preserved by Roman tradition, disappeared at an early date. Some of the inhabitants of these towns were transferred to Rome. Those belonging to the local nobility may perhaps have been incorporated in the Roman Patriciate, while the others were absorbed by the Plebs, and this influx still further upset the balance between the two rival elements of the Roman people.

Nor was this all. Economically, perhaps the most

[1] **LXVIII**, p. 48.

serious result of the conquests of Rome in Italy was the change which took place in the ownership of land, in its character, its organization, and its distribution. By the ancient laws of war, a conquered city ceased to belong to itself and became in every respect the property of its conquerors. Its territory legally became their absolute property. In practice, Rome did not confiscate all the land conquered; indeed, she left the greater part of it to the former owners, under certain legal and fiscal conditions. The rest she divided into several parts. Part she sold or distributed to individuals who enjoyed the full ownership of it; this became *agri privati*, over which the Roman State no longer had any right. Another part she leased out, usually by auction, for an annual rent; this land continued to be the legal property of the State, the farmers only having the *possessio* of it. Another part was conceded to the first occupier, who had to pay the State an annual rent like the farmers of the previous category. The lands of these two last classes did not become *agri privati*, but remained part of the *ager publicus*. Now, it came about, by the mere force of circumstances, that sold land, leased land, and land conceded to the first occupier were all alike taken up by wealthy Patricians or Plebeians who had made money. The small and middling landowners and free tenants, being ruined by the war as we have seen above, could neither buy the first kind of land nor hire the second, and the land which was conceded to the first occupier almost all fell into the hands of the rich, who alone had the equipment needed to make it productive.

This was the origin of the land question, which was as serious and acute as that of debt. Whatever one may consider to be the date of the Licinio-Sextian Laws, placed by Roman tradition in 367 B.C., whatever one may think of their authenticity as measures taken about the middle of the fourth century B.C., the fact that such laws were passed proves, first that the mass of the poor suffered cruelly from the greed of the rich and the practice of usury, and secondly that the Patricians and wealthy Plebeians had appropriated enormous areas, each over 500 *jugera*, of the public land. In consequence, while for various reasons small and medium properties became fewer, big properties increased and invaded the *ager Romanus*. The distribution of individual allotments of land in the Roman or Latin colonies was an insufficient palliative to a situation which was full of danger for the future, especially since the laws passed in an attempt to solve the two problems of land and debt were usually either openly violated or quietly evaded.[1]

About the industry of Rome at this time, we have less information than about the development of landed property and its consequences to agriculture. It is likely that the advance of the power of Rome and the growth of the city helped to attract craftsmen and to develop industrial work. The metal-workers, weavers, potters, goldsmiths, armourers, and manufacturers of toilet articles seem to have constituted the busiest trades, and doubtless already made up a large part of the Plebs.[2] The institution of the worship of Minerva, the great goddess of workers, on the Aventine is evidence of this economic movement and definitely shows its character.[3] The Vicus Tuscus of Rome, as the Velabrum was called, is believed to have been originally inhabited by craftsmen from Etruria. Industry remained stationary, however, and played no very great part in the economic life of Rome from the fifth century to the third.[4]

Trade, on the contrary, made marked progress. Thanks to her geographical position, Rome became the chief market of all central Italy, and she was near enough to the sea to be in touch with the marine trade which then flourished in the Western Mediterranean. Grain was imported in great quantities at an early date. There is evidence for this in the introduction into the Roman religion of the Greek triad of Demeter, Dionysos, and Core, under the old Italic names of Ceres, Liber, and Libera, which Roman tradition placed in the first years of the fifth century. "It went together with the development of a very active trade in corn between Rome and the regions of the south of Italy, Campania, and Sicily. . . . Native production was not sufficient to meet all needs, and the extra flour required to feed the citizens had to be procured outside Latium. Consequently, a movement of foodstuffs, chiefly corn, set up from the neighbouring districts to the mouth of the Tiber."[5] No less significant is the appearance in Rome of Mercury, the god of traders

[1] **LXVIII**, pp. 80 *ff*. [2] **LXIX**, pp. 29, 39. [3] **LXXV**, pp. 184 *ff*.
[4] **LXIX**, pp. 62 *ff*. [5] **LXXV**, pp. 143-44.

who hailed, like the triad of Ceres, Liber, and Libera, from the Hellenized parts of southern Italy.[1] The trade of Rome and Latium with Great Greece seems to have been fed to some small extent by the products of the ceramic and metal-chasing industries. It is possible that Rome exported to the Greek world copper which she bought in Etruria and wool and hides which she obtained from the pastoral peoples of the Apennines.

Although we know little of the details of Roman trade in this period, we can at least obtain an idea of its expansion, thanks to certain facts and several documents of the greatest interest. Even if we admit that the Roman colony of Ostia was not founded before the middle of the fourth century, we may at least be allowed to suppose that an inhabited centre previously existed at that spot, and therefore that marine trade already made use of the mouth of the Tiber. The text of the first commercial treaty concluded between Rome and Carthage mentions, among the towns subject to Rome, Antium, Circeii, and Tarracina, and Polybios, who has preserved the text for us, declares that the treaty was signed "under the first Consuls of the Republic," that is, in 509-508 B.C. Modern historians do not agree as to how much trust should here be placed in Polybios. Messrs. Gsell and Frank accept his date;[2] M. Homo holds that he has made a mistake, and that this first treaty is not earlier than the middle of the fourth century.[3] Even if we adopt this latter view, it remains certain that as early as 350 Rome had control of the ports situated along the coast of Latium and the Volscian country. This contact with the sea and this control of several ports made it necessary for Rome to possess a merchant navy, and indeed this fact emerges from the very wording of the two treaties which she struck with Carthage:

> "Neither the Romans nor their allies shall sail beyond" (*i.e.*, west of) "the Fair Promontory. . . . If Romans land in the parts of Sicily belonging to Carthage. . . . In Sardinia and in Libya, no Roman shall do trade, or found cities, (or land) longer than is needed to take on provisions and repair his ship. If he is driven there by a storm, he shall depart within five days," and so on.[4]

At the end of the fourth century, Rome made a treaty with the great Greek colony of Tarentum; "an article in it

[1] **LXXV**, pp. 181 *ff*.
[2] See above, pp. 201-2; **LXIX**, p. 30.
[3] **LXXIII**, English p. 176.
[4] *Cf*. above, pp. 201-2.

formally forbade the Roman fleets to pass the Lacinian promontory."[1] So Roman vessels were already sailing about the Tyrrhenian Sea and going through the Strait of Messina.

Inside Italy itself, Roman trade had extended into one district after another with her political dominion. It followed the chief roads which ran south, like the Appian Way, which was built as early as the end of the fourth century, or east and north-east, like the Latin and Flaminian Ways, or north-west, like the Aurelian Way. These roads, on which many colonies had been founded, were of as great value economically as strategically.

Another characteristic sign of the economic development of Rome from the middle of the fourth century onwards was the appearance at that time of an official coinage. Hitherto, since the primitive days of barter, the Romans had employed the method of weighing metal. *Æs rude* consisted of lumps of copper unadorned with any effigy or stamp; sales were said to be effected *per æs et libram*, "by copper and scales." A method of exchange as rudimentary as this became quite insufficient when the economic life of Rome depended on more highly civilized peoples like the Greeks, whose colonies in Great Greece and Sicily had long used coins which were convenient and much appreciated. Accordingly, the Roman State struck bronze and silver coins. The first bronze coins were of various values—the primitive *asses* which weighed a Roman pound and coins representing multiples and fractions of the pound. The silver coin, which was at first intended solely for military and commercial purposes in southern Italy, was a two-drachma piece of the type then current in Campania, and was minted at Capua. These Roman didrachms, which weighed about 7·5 grammes of silver, could doubtless be exchanged against three copper *asses* each weighing a Roman pound—*i.e.*, 327 grammes. The disadvantages of this partly dual system were perceived fairly early, and after various experiments, into the details of which we need not enter, and reforms necessitated by fluctuations in the market price of copper in Rome, a new system was adopted in 269 B.C. The Roman pound of 327 grammes was taken as a basis, with its division into 12 ounces and 288 scruples. The copper unit was the *as* of 2 ounces or

[1] LXXIII, English p. 203.

48 scruples; the silver unit was the denarius of 4 scruples, which roughly corresponded to the Athenian drachma. A silver denarius was worth 10 copper *asses*. Copper coins representing fractions of the *as* were struck, and also a small silver coin equal to a quarter of a denarius, called the sesterce.

Rome now possessed an instrument of exchange which greatly promoted her trade. The innovation was especially important to her economic life in that it coincided with the extension of her territorial and political power, principally in the south of Italy, that is, in the region which contained the busiest markets with which she dealt. Besides, the appearance of a coinage which was easy to handle and easy to exchange with current Greek coins made a great difference to commercial transactions. Sales, purchases, and transit operations were greatly stimulated; there were even financial crises, created by the much more active circulation of coin.

But these were only passing troubles, mainly due to the inexperience of the magistrates of the Republic in financial matters. The great economic consequence was the development of movable wealth in Rome. Formerly, when the census was taken and the citizens were divided into classes and tribes, landed property alone had been taken into account. In 312, when Appius Claudius was elected Censor, he took statistics of movable capital, and his successors did not dare to undo his work. It was an absolute revolution in law; it was nothing less than "the entrance of movable wealth into political life."[1]

Another result of the conquests of Rome, no less fruitful from the economic point of view, was the fiscal organization of the *ager publicus*. In theory, the territory of the defeated and conquered cities belonged to Rome. We have seen above that part of this territory was sold, leased out, or conceded to the first occupier, but it was only to Roman citizens. Another part of the territory could be distributed to colonists, Roman or Latin. Another part, finally, was left to the former owners, grouped in a municipium, a prefecture, or an allied city, but on condition that they paid tribute (*vectigal*) to Rome. Moreover, to meet the often considerable expenses imposed on Rome by the frequent, almost uninterrupted wars in which she conquered the peninsula, even Roman citizens

[1] **LXVIII**, pp. 109 *ff.*, 118.

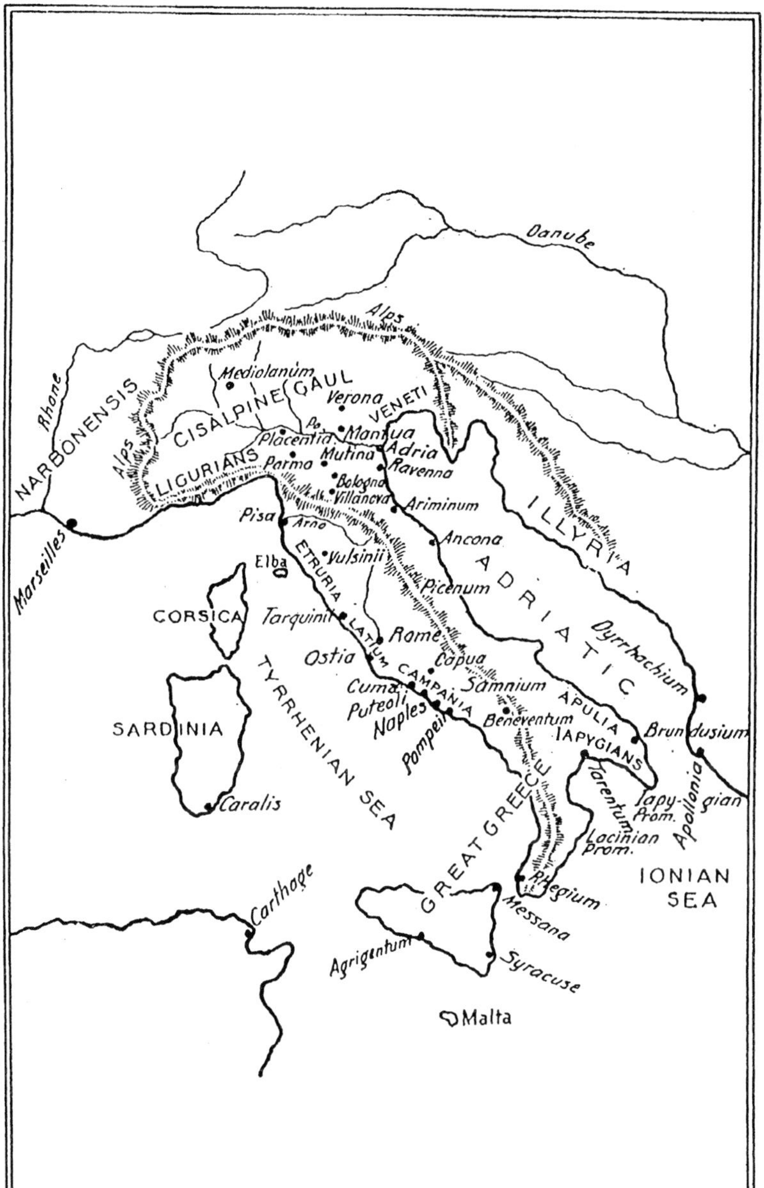

MAP V.—ITALY UNDER THE ROMAN REPUBLIC.

had to pay a tax, the amount of which was assessed according to their total resources. For a long time the authorities based their assessment on landed property alone; movable wealth was included from the censorship of Appius Claudius in 312 B.C. Lastly, customs are proved to have been established on the frontiers of Roman territory, perhaps in the Royal period and at any rate in the first centuries of the Republic.[1] Roman citizens, Latins, and Italian subjects and allies were all alike subject to these customs-dues or *portoria*. To collect all these direct and indirect taxes, the proceeds of the sale and hire of land, the rent from conceded land, and customs-duties, the State set up a fiscal organization in which the system of "farming" figured largely. This was the beginning of an evolution which was at first mainly financial, but afterwards had a considerable influence on the economic life of Rome and the whole Roman world.

We have now reached the moment at which Rome, the capital of an Italy united under her rule, was about to enter into conflict with Carthage. The stake for which they would fight was the supremacy of the Western Mediterranean. At the same time, Rome began to interfere in Greece and the East. For about two hundred years, in a series of wars in which she would not meet with uninterrupted success but would always obtain the final victory, she would collect the whole ancient world under her sway, and would establish the unity of the Mediterranean for her own benefit. From being urban and regional, her economic life would become, if one may use the words regarding antiquity, international and world-wide. Here, again, the economic history of the Roman State would be determined, like her social evolution, by foreign affairs.

[1] **LXXIII**, English pp. 238-39.

CHAPTER IV

THE ECONOMIC LIFE OF ROME FROM THE BEGINNING OF THE PUNIC WARS TO THE DEATH OF AUGUSTUS (264 B.C.–A.D. 14)

DURING the period of nearly three hundred years which lies between the beginning of the Punic Wars—that is, the first attempt of Rome to extend her domain outside the peninsula of Italy—and the death of Augustus—that is, the time when the Empire was constituted in its chief territorial limits and in the general form which it would more or less retain down to the barbarian invasions—the Roman State absorbed, one after another, the various Mediterranean countries in which economic life had developed at various rates and under different conditions. Her victory over Carthage made Rome mistress of the Western Mediterranean; the conquest of Greece and the defeat of the Oriental sovereigns, Antiochos, Mithradates, and Cleopatra later, gave her the dominion of the eastern seas; the submission of Gaul up to the Rhine and the occupation and organization of the vast regions on the right bank of the Danube from the Lake of Constance to the Black Sea carried her economic influence with her political power into the heart of Central Europe. When that achievement was completed (to be rounded off later in Europe by the annexation of a great part of Britain, the Agri Decumates between the upper Rhine and upper Danube, and Dacia, and in Asia by that of Mesopotamia and Arabia), the ancient world, once divided and tugged about between frequently warring states and kings, found itself collected in a single political organism, under the direction of a powerful mistress who was able to maintain that new unity, strong enough to allow free play to the economic forces working within the limits of the Empire, and also strong enough and prosperous enough to attract to herself or to reach the resources of vast countries outside, such as India, China, and Central and East Africa.

An evolution which produced such consequences as these could not, so long as it was going on, fail to affect the economic life of Rome herself. Before it reached that state of equilibrium which it enjoyed for at least the first two centuries of the Empire, that economic life went through a series of stages, which were determined by the various fortunes of the wars waged by Rome and the results of her final victories. One of these results, perhaps the most important from the point of view which concerns us here, was the change which took place in the character of the city itself. It ceased to be merely the capital of Rome and Latium; gradually, as its sway extended over West and East, it became a cosmopolitan caravanserai, into which men swarmed from all parts, chiefly Greeks and Orientals. It was invaded by a flood of foreigners, free and slave. The terrible holes made in the citizen body by frequent wars seemed to fill up rapidly; there might even be the illusion of a continual increase, since, according to Beloch, the lists made by the Censors rose in less than a hundred years from 214,000 entries in 203 to 394,000 in 125. But the men whose names filled these lists were very different. Freedom of foreign origin took a bigger and bigger place in them. We know the famous retorts of Scipio Æmilianus. "One day when he was interrupted by the populace, he hurled his contempt at them, 'Silence, you whom Italy does not own as her children!'; and when the shouts redoubled, 'I brought you here loaded with chains; you will not frighten me now that your chains are off.'"[1]

Such a racial and social transformation as this could not leave the economic life of the city unaffected. And indeed, in the course of the last centuries of the Republic, great changes took place in the agriculture of Rome and Italy, in industry, in trade, and in the fiscal and financial organization of the State.

I

The Agriculture of Rome and Italy

The causes which had begun to influence the conditions and character of agriculture between the fifth and third centuries B.C. continued to influence them, with much greater

[1] **LXVIII**, p. 166.

intensity, during the wars in which Rome conquered the whole Mediterranean. The burden of those wars bore on the small and medium landowner and free tenant far more heavily than that of the wars against the peoples of Italy had done. Military service no longer kept the citizen away from his fields for a few months in the year only, but for long years. The first Punic War lasted over twenty years, the second, eighteen, and hardly was the conflict with Carthage over when Rome commenced her expeditions in the East, against Macedon, against Syria, against Greece, to say nothing of the wars which she started or had to maintain in northern Italy, Spain, and southern Gaul. "To hold her own on every side," M. Bloch writes, "in Cisalpine Gaul and Liguria, in Spain, in Corsica, in Sardinia, in Macedonia, in Africa, she needed at least eight legions, levied almost regularly every year, not including those maintained in arms when circumstances required and not including men sent out to fill up gaps or retained to give a stiffening to new legions of recruits. . . . The consumption of men was frightful."[1] In the course of these wars, many legionaries fell on the battlefield; the defeats of Trasimene and Cannæ were absolute hecatombs of Roman soldiers. Those who came through alive had not always left sufficient workers behind them for the normal and complete farming of their modest estates. The mere fact of continuous military activity was already a cause of decline and ruin to small and medium-sized property. That decline and ruin were aggravated by the ravaging which Italy underwent during the second Punic War. "For more than twelve years the battle lines swept back and forth over the villages and fields of central and southern Italy. Cities surrendered to the seemingly stronger contestant for self-protection, only to be sacked in vengeance when captured by the other. Whatever contestant retreated, grainfields were burned for military reasons, vineyards and orchards cut, and the cattle driven off. The inhabitants who escaped scattered to the four winds, many abandoning Italy permanently for Greece. Many of the famous cities of Magna Græcia came out of the war with a few hundred famine-ridden weaklings huddling together along the ruins of the city walls."[2]

[1] *Ibid.*, pp. 147-48. [2] **LXIX**, p. 89.

The middle class, which supplied the agriculture of Rome and Italy with its most intelligent directors and its hardest workers, was decimated and diverted from its ancient task. The ground itself, the plantations and crops which required prolonged and regular attention, had gone back badly. Yet the restoration of the old rural economy would not have been beyond the strength and courage of the Roman people and the inhabitants of Italy if other causes, economic and social, had not stood in the way.

The wars undertaken by Rome did not bring only damage and loss. Her victories brought a considerable extension of the *ager publicus*. We have seen above how confiscated land was sold or leased out or left to the first occupier in return for a rent, and we have seen why, almost inevitably, it was only the rich, the great landowners, who could buy, rent, or occupy these various classes of land. This situation, which was bad enough in the fourth century, grew still worse in the following centuries. " It was the rich who benefited by the development of the *ager publicus*. They alone had the capital needed to purchase the right to exploit the forests and to collect the tax on pastures. They alone could obtain the labour needed to clear waste land. The land of which they took possession was not only that which they themselves had put or were putting in a state of cultivation. It was all the land which they claimed to be able to clear in a more or less remote future, so that there was no longer any limit to their encroachment. Of the limitation to 500 *jugera* there was no longer any question. The law which had established that maximum had long fallen into abeyance. Really, the occupier was not the owner; he had ' possession ' (*possessio*), not ownership (*dominium*). The owner was the Roman people, the State, whose right could not be taken away. Against any other than the State, the possessor was guaranteed; against the State, he was helpless. The State secured him in the enjoyment of his property, but, by the rent which it took, it asserted the perpetuity of its right. That was the theory. In practice, the State came in the end not to exact the rent. The possessor succeeded in emancipating himself from it, as he had emancipated himself from the restrictions imposed by the Licinian Law. Then, when the State allowed its title to lapse, he began to ignore it; he regarded it as null and void, he regarded himself as a true owner, and neither he nor the State itself made any distinction between his 'possessions' and his private property."[1]

Furthermore, the big estates formed in this way lay alongside of small properties, or might even surround them. As happens in such cases, the owner of the big estate wanted to round it off by obtaining the adjoining land. "Failing a purchase by private treaty, there was a war of trickery, legal expropriation, and violent eviction without formality. . . . 'The rich,' Appian tells us, 'bought or took by force the little inheritances of their poor neighbours.' Plutarch tells the same story. The absence of the head of the family on military service made such usurpation all the easier. 'Their parents and children, if they had some powerful neighbour, were driven by him from their homes'; so Sallust speaks. These evictions became a literary theme exploited by rhetoricians and poets."[2]

So, as a result of various causes and by different methods, there grew up the huge *latifundia*, of which Pliny the Elder says, in a much quoted passage, that they destroyed Italy. Here and there, no doubt, chiefly in the mountainous country of central Italy, in the valleys and on the slopes of the Apennines, small property did not vanish altogether; but only fragments of it remained. Intelligent or ambitious reformers attempted, some by creating colonies, like C. Flaminius in 232 and Caius Gracchus in 123-122, and others by agrarian laws, like Tiberius Gracchus in 133, to remedy the economic evils which this predominance of the big estate brought on Rome and all Italy. None of them succeeded.

Another equally serious development which helped to alter the character of landed property was the unprecedented extension of slavery. Slave labour took the place of free labour on most farms. Almost every year, Rome and Italy saw an influx of prisoners of war, who had been taken by the legions on every battlefield in the West and in the East. These prisoners were sold as slaves. We hear of a campaign in consequence of which 150,000 human beings were reduced to bondage. Moreover, the Eastern Mediterranean was the scene of a regular slave-trade. One of the profits of piracy, then in its heyday, was the sale of men and women kidnapped

[1] **LXVIII**, pp. 153 *ff*.

[2] *Ibid*., pp. 158 *ff*.

in Greece, Asia Minor, Syria, and other places. The great slave-market was the island of Delos, where, on certain days, according to Strabo, over 10,000 unfortunates were put up to auction. At prices varying between £20 and £60 a head according to the age, strength, and capacity of the individual, the great landowners of Rome and Italy could procure abundant and certain labour, which, by unions between slaves on the same estate and the birth of children who were slaves from the moment they opened their eyes, increased unceasingly. In spite of the cost of keep, slave labour was cheap. It did not cause free labour to disappear entirely. In the last centuries of the Republic there were still agricultural day-workers and tenants who paid a proportion of the produce. But on most estates of any size the work was done by slaves.

The predominance of the big estate and the extension of slave labour—these were two economic facts, contemporary and parallel, which did not fail to influence each other mutually. One cannot say that one was the cause of the other; but one cannot deny that the system of the *latifundia* was encouraged by the influx of slaves and that, on the other hand, the preference given to slave labour was a consequence of the creation of those huge estates.

However that may have been, the new system of landed property meant, for Rome and Italy, the transformation of agriculture and agricultural science. Corn-growing, which for long had been the chief occupation of the Italian peasantry, declined steadily. It must also be recognized that the conquest and annexation of Sicily and Sardinia, and later of the territory of Carthage, Asia Minor, and finally Egypt, encouraged the importation of corn on a large scale, against which the Italian grower could only fight with great difficulty. The countries which were made into provinces became the granaries of Rome. The big landowners would no longer put their estates under corn. They preferred growing fruit and trees, among which the olive and vine took the first place. Round the towns and in especially favoured districts like Campania, orchards and gardens developed at the same time as olive-groves and vineyards. Vegetables were grown in the shade of the fruit-trees or between the rows of the vine-stocks.

But it was above all stock-breeding which took the place of corn-growing, sheep, pigs, and horses being chiefly raised. Vast areas, left waste or simply turned into pastures, were given over to immense flocks of sheep. Stock-owners already practised the system of keeping their beasts in the plains during the winter and taking them up to the hills, the southern Apennines or the Abruzzi, in the first days of summer. In addition, as luxurious tastes spread, the rich made aviaries, poultry-yards, fish-ponds, and game-parks on their estates.

The information which the ancient authors give us does not allow us to state, with the exactitude of modern statistics, either the average extent of the big estates which were formed at that time or the comparative areas given to the different kinds of farming. But here and there we find general indications which are useful, and archæology has supplied valuable data regarding the results of the process through which landed property was then passing. "Cato, in his treatise on agriculture, is thinking of an estate comprising 240 *jugera* (150 acres) of olive-grove and 100 *jugera* (62 acres) of vineyard. That is a medium-sized estate; but he says nothing of arable and pasture. Cato lived from 234 to 149. Varro, who wrote his *De Re Rustica* in 37, addresses himself to big proprietors as much as to medium ones."[1] The small property does not seem to have interested the agricultural writers any longer. This is the order in which Cato gives the various crops of an estate in respect of their importance: (i) vines; (ii) garden produce; (iii) willows; (iv) olives; (v) meadows; (vi) corn; (vii, viii, ix) various kinds of wood—fuel, fruit-trees, oak. Pliny says that Cato placed stock-breeding first; he probably did so in a work now lost.[2]

The falling off of crop-growing and the advance of stock-breeding also led to the neglect and abandonment of vast areas which had once been productive. Plutarch relates that the idea of the agrarian reform occurred to Tiberius Gracchus "when, on the way from Rome to Numantia through Etruria, he saw the country deserted, the only field-workers and herdsmen being foreign slaves and barbarians."[3] It is very probable that in Etruria, the Roman

[1] **LXVIII**, p. 156.
[2] **LXIX**, p. 97 n. 13.
[3] *Tib. Gracchus*, 8.

Campagna, and the Pontine Marshes the old drainage-works at this time fell into ruins, and that it is from this disappearance of regular, methodical cultivation, this interruption of human labour, of the continual effort of the tillers of the soil, that one should date the return of unhealthiness and malaria to the offensive. Not nature is to blame here, but man himself.

This transformation of rural economy led in the end to a social phenomenon of great importance in the history of Rome. The countryside was deserted by most of the small landowners and free tenants, who came crowding into the towns, and especially to Rome. This influx of poor people, ruined by the new conditions of farming, enormously increased the mass of the Plebeians, which was at the same time being swelled by increasing throngs of freedmen of foreign origin, Greek and Oriental. There was no longer a balance in the citizen body between the rural elements, which had once been hard-working, productive, and level-headed, and the urban proletariate, which grew more and more accustomed to call upon the State, the magistrates, candidates for office, and ambitious men of every kind for its food and amusements—*panem et circenses.* Rome ceased to be the capital of an essentially agricultural people, whose wealth was mainly based on landed property and agricultural resources; it became a turbulent agglomeration, in which industry, trade, and money-dealing assumed an importance hitherto unknown.

II

Industry in Rome and Italy

In his work on the *Economic History of Rome to the End of the Republic*, Mr. Tenney Frank, after observing that no appreciable progress was made in the domain of industry in the fourth century B.C., adds: " In the two succeeding centuries we do not find evidence of any marked change in the nature of production at Rome. Doubtless the amount of ordinary ware produced at home increased with the growth of the city . . . but of goods worthy of export we do not hear. The only difference now is that work previously performed by free labour began in the second century to fall into the hands of slaves."[1] As for the picture which he draws of industry in the first century B.C., it seems to me to apply to the beginning of the Empire much rather than to the end of the Republic, whether it is a question of Rome and Italy in general or of Pompeii in particular.[2]

The ancient authors, it is true, do not give much information on the subject, and there is no great abundance of archæological or epigraphic documents to make up for their silence. Direct evidence is rather scanty than otherwise. Lacking such evidence, what we know of the general history of Rome and Italy allows us to obtain a glimpse, by deduction, of the intensity and character of industry at that time.

In Rome itself, it is incontestable that the various trades which produced the necessities of daily life must have increased their output in order to satisfy the needs of an urban population which grew bigger every year. All these townsmen, who numbered hundreds of thousands, needed bread and other things to feed them, cloth and garments to clothe them, instruments, tools, furniture, and a thousand different articles for their work or for the adornment of their houses, whether rich or poor, magnificent or squalid. That commerce was able to supply some of these things is possible, and we shall see so later on. Nevertheless, it will surely be admitted that most of them were made on the spot, and that in consequence many industries developed in Rome itself and in the chief towns of Italy. Besides, it should not be forgotten that in the last centuries of the Republic the city grew much larger and more ornamental; houses had to be built for the masses crowded within its walls, and many public monuments were erected. Lastly, all through this time the military and naval power of Rome developed greatly, and one can well imagine that the manufacture of munitions and shipbuilding were greatly promoted thereby.

There is no doubt that the making of food, textiles, metal goods, and pottery, the industries which use leather and wood, the manufacture of all engines of war, offensive and defensive, and the building of war navies and merchant navies had a very high place in the economy of Rome in the last centuries of the Republic. Perhaps one or other of these industries was still partly of a domestic nature. In many

[1] **LXIX**, pp. 102 *ff*. [2] Chaps. xi-xii, pp. 165 *ff*., 190 *ff*.

houses, chiefly in the country and sometimes also in town, the bread was still kneaded and baked for the use of the family, and wool was spun and woven. On the big farms, where the numerous slaves had to be kept busy all the year round, baskets were made of osiers grown on the estate, and where the soil provided a good clay big crocks were made to hold corn, oil, and wine. Often an olive-grower would have his own oil-press. But these were exceptions. An organization of this kind, in the then social state of Rome and Italy, could only suit a few industries. All or almost all were specialized. Even if their output was absorbed by local or regional consumption, and could not, as Mr. Frank says, feed an export trade, the market open to them in Rome and Italy was extensive and important enough to keep them busy and prosperous.

About this industrial development we know few definite details. Rome did not allow the mines and quarries which she had secured by her conquests in Sicily, Spain, and Macedonia to lie idle; she continued to extract iron-ore from the mines in northern Etruria; later she had also the metallic wealth of Gaul and Noricum. Thanks to these new and plentiful supplies of raw materials, the metal industry developed greatly, if not in Rome itself, where we do not hear of foundries or workshops, at least in Etruria and later in Campania, where several cities, Cales, Capua, and above all Puteoli, became very active centres of production. Diodoros[1] tells us that at Puteoli there were many blacksmiths, making weapons, spades, picks, scythes, sickles, and tools and implements of all kinds.[2] Capua seems to have been the centre of the copper and bronze industries, which had doubtless been started in the time of the Etruscan domination. The workshops of this city turned out masses of pails, dishes, ladles, bowls, pots, saucepans, and other kitchen gear, to say nothing of countless artistic objects, such as statues, busts, and statuettes similar to those dug up in the ruins of Herculaneum and Pompeii.[3] In addition to these stationary industries, every army had attached to it bodies of workmen who repaired swords, shields, helmets, and the engines which were the artillery of the day, and

[1] Diod. Sic., v. 13.

[2] **LXIX**, p. 178.

[3] *Ibid.*, pp. 181 *ff.*

supplied the combatants with the spear-heads of which there was a great expenditure in every campaign. In Rome itself there were smaller workshops, where the legionary obtained his arms before proceeding to the field.[1]

With the building of the older aqueducts which brought Rome the water needed by a big urban agglomeration a manufacture of lead pipes sprang up, but this industry did not attain its full activity until the Empire.

The working of the precious metals was as flourishing as that of the common metals. The luxurious tastes which were gradually spreading to all classes of Roman society gave rise to a demand for jewels, gold and silver rings, ear-drops, and trinkets adorned with gems. Goldsmiths and jewellers were numerous in Rome at the end of the Republic; their shops, which were also their workshops, stood chiefly on the Sacred Way.[2] The intaglio-cutters found plenty of custom among the wealthy Romans, all of whom wore seals in their rings.[3]

Pottery benefited no less than metal-working by the advance of the power of Rome. On public buildings, no doubt, under the influence of Greek art, stone and marble took the place of the terra-cotta which the Romans had learned to use from the Etruscans. But for private dwellings, if we are to believe Augustus when he says in his will that he found Rome brick and left it marble, the making of common bricks and tiles must have had a big place in the building-trade. There were plenty of beds of excellent clay in the environs of the city; the alluvia of the Anio and Tiber had deposited geological strata in which the calcareous elements from the slopes of the Apennines were combined with the volcanic *pozzuolana* of Latium.[4] Besides the bricks and tiles used for buildings, the Roman potters, long before the Empire, produced conduits and waterpipes for private houses and estates.

The importance of public buildings and other public works and, therefore, the activity displayed by the building-industries at that time, are attested by the many contracts given out by the Censors every five years. Livy mentions them frequently; hardly one space of five years goes by but

[1] *Ibid.*, pp. 180 *ff.*
[2] *Ibid.*, p. 187.
[3] *Ibid.*, pp. 118 *ff.*
[4] *Ibid.*, p. 174.

he speaks of one or more. Among the Roman monuments which date from this period, we may mention the Porcian, Æmilian, Sempronian, and Opimian Basilicas, built round the Forum, the second temple on the Capitol (that which was built by Sulla and Catulus), the Tabularium, the Theatre of Pompey, etc.[1] It was about the end of the Republic, too, that the construction of the network of Roman roads over Italy and the provinces was undertaken. One may also assume that much work was done on the Italian ports, at Puteoli, at Brundisium, at Ancona, to say nothing of Cæsar's plans for restoring Corinth and Carthage and cutting a canal through the Isthmus of Corinth.

Of the textile industries in the last centuries before the Christian era we know next to nothing, so far as Rome and Italy are concerned. One or two details only emerge from the obscurity. Cato the Elder lamented that the Roman matron of his day had lost the habit of spinning wool. No doubt, she made her slaves do it, but one may suppose that spinning and weaving had become specialized trades. Besides, the poor, who had to work for their living and could not afford even one slave, would have found it exceedingly difficult to make their own garments, however primitive we may suppose them to have been. The dressmaking industry had become indispensable. So, too, the passing in 217 of a Metilian Law regarding Fullers, which laid down regulations for the organization of that trade, is sufficient to prove that the manufacture and treatment of cloth were specialized by the second half of the third century.

Rare and scattered as these indications are, they suffice to throw light on the activity of industrial life in Rome and Italy between the Punic Wars and the establishment of the Empire. The actual organization of that industrial life, however, deserves attention. Here, as in the case of agriculture, one of the most important facts is the part played by slavery.

In the town as in the country, in industry no less than in agriculture, slave labour took an unprecedented place. Wealthy Romans set up workshops of slaves under the management either of a slave or of a freedman. Some of

[1] Platner, *Topography and Monuments of Ancient Rome*, pp. 169, 300, 306, 366.

these workshops provided the needs of the master's house; others manufactured for sale. In the latter case, one may safely say that much capital must often have been spent on organization and the daily costs of production. In a humbler sphere, a craftsman of free birth would employ a slave or two in his workshop, which was also his saleroom. At this time the position of the freedmen, as captains of industry, became very important. The man whom the moneyed Roman placed in charge of his workshop of slaves was frequently one of his former slaves, whom he had himself freed. Other freedmen would start a business of their own, with such money as they had got together. One should not forget, to explain this general character of the organization of industry in Rome, that manual labour was held in very low esteem. Philosophers, like Cicero and Poseidonios, did not disguise their contempt for the "working-class dregs."[1]

But it would be too much to say that there was no free labour in Roman society in the time of the Scipios, the Gracchi, and Cæsar. Trade-guilds already existed. I have mentioned a law regarding fullers; other guilds are known in the last centuries of the Republic—those of the goldsmiths (*aurifices*), building-workers in general (*fabri*), potters (*fictores*), rope-makers (*restiones*), stone-sawyers (*sectores serrarii*), etc.[2] Varro speaks of blacksmiths who went about the country, from farm to farm;[3] they doubtless repaired damaged farm-gear and made certain tools on the spot. The fact that they moved about shows that they were workers of free estate.

We need not dwell further on the economic and social organization of labour in the Roman Republic. The great majority of our documents on the subject are no earlier than the Empire, and we shall discuss it presently.

III

Trade in Rome and Italy

About Roman trade in the last centuries of the Republic, Mr. Tenney Frank seems to give utterance to an opinion as disputable as that maintained by him regarding the position

[1] **LXIX**, p. 266. [2] **CIV**, vol. i, pp. 87 *ff.*
[3] **LXIX**, p. 180 n. 15.

of industry in the same period. That opinion appears to be summed up in this typical sentence: "Indeed the ancient world has no record of any state of importance so unconcerned about its commerce as was the Roman Republic."[1]

No doubt, Mr. Frank means to say that the actual town of Rome did not become a very busy commercial centre in the last centuries of the Republic, and that in particular it could not furnish material for a big export trade. But, for one thing, it is an unwarranted limitation of the subject to confine the study of it to the observation of the business which went on in Rome itself. The economic activity of Rome extended to the whole of Italy, then to the Western Mediterranean, and finally to Greece and the East. The economic history of the Roman State can no more be confined to that of the town of Rome than the political history of that State can exclude that of Italy and the provinces. The chief traders of the time may have been Italians of Great Greece, or Greeks, or Orientals, and in Rome itself the freedmen, usually of foreign origin, may have had a preponderant place in business; the fact remains that all this movement of business was closely connected with the economic life of Rome and that it was a direct consequence of the advance of the power of Rome, and not to be separated from it. For another thing, it is, if not incorrect, at least an exaggeration to say that in Rome itself trade had made no serious progress. The imports, which became necessary with the mere increase in the size of the town, and had to satisfy larger needs every year, should not be treated as negligible in an appreciation of the commercial activity of a state. Even if in Rome they were not counterbalanced by corresponding exports, the balance of trade was established in another form and by other methods.

The commercial prosperity of Rome at the end of the Republic has, moreover, left material witnesses in enlargements and embellishments of the quarter of the Aventine and the plain below it, where the river-port of the Tiber was situated, the true emporium of the city. "Dionysios of Halicarnassos . . . in the last years of the Republic, laid weight on the contrast presented by the Aventine as it had been, covered with woods and uninhabited, with what one

[1] **LXIX**, p. 110.

saw there in his lifetime. The old clumps of trees, the little laurel-wood, the fields, the great empty spaces, had all disappeared, and had given place to houses crowded one against another and cramped within an uninterrupted line of temples—*νῦν δ'οἰκιῶν ἐστι πλήρης ἅπας*. . . .

"The hill was still what its geographical position predisposed it to be, and what it had been for centuries—the commercial centre of Rome. It was thanks to the undertakings of the business men and their preponderance, which increased with their success and their profits, that the Aventine, after Hannibal's time, developed, perfected the mechanism of its daily life, and secured more well-being and comfort for itself; just as it had been through them, through the effort which they had made to secure the transit trade of Rome, that it had risen from its old position of inferiority and had begun to count for something in the destinies of the city."[1]

The river-port itself was extended and improved more than once. New wharves, porticoes, warehouses, stairs, and stores were built in the course of the second century B.C. M. Merlin rightly concludes that all these structures were intended "to receive the goods which came pouring into Rome from every place in the world."[2]

At the same time as the Aventine and the emporium were being transformed to suit the commercial progress of the city, the prosperity of Puteoli, which was before Ostia the real port of Rome on the Tyrrhenian Sea, pointed to the extension of Italian business in its connexions with East and West.[3]

What were the elements of this traffic? Rome and Italy imported raw materials, foodstuffs, and manufactured goods, objects of luxury and of art. From Gaul and Spain they obtained metals—copper, lead, silver, tin—raw wool, hides, and leather. Asia, Pontos, and Syria sent them timber. Egypt passed on the ivory which she got from Ethiopia, and already Gaza was sending the spices, frankincense, and myrrh brought by the caravans from Arabia. Sicily and the fertile plains of Asia Minor sold them corn. Cargoes of salt fish came from the Euxine and Spain, salt goods and

1 **LXXV**, pp. 250, 252. 2 *Ibid.*, p. 251.
3 C. Dubois, *Pouzzoles antique*, pp. 64 *ff.*; *cf.* **LXIX**, pp. 248 *ff.*

cheese from Gaul, and wines of renowned brands from various regions of the Mediterranean. The wealthy Romans eagerly bought the fine tissues, carpets, and glassware made by the towns of Asia Minor, Phœnicia, and Syria and by Alexandria in Egypt. They sent to Greece for artistic objects; Cicero asked Atticus to buy and send him statues for the adornment of his villas at Tusculum and Formiæ.[1] The ship laden with bronzes and marbles which sank off the Tunisian coast, whose wreck, deep under the water, has been explored under the direction of M. Merlin, was doubtless carrying her cargo of works of art to Italy when she was driven by storms into those African waters.[2] The slave-trade was not the least prosperous of all these activities; the principal markets were in the East, but Gaul and the countries of the North also sold that deplorable human merchandise.

Exports were far from balancing these varied and plentiful imports. Rome and Latium could not supply a return cargo of any value to the ships which went up the Tiber to Rome or stopped at Ostia. Only some regions of Italy sent a few goods abroad. Campania exported bronze and iron goods and pottery, and the Po valley pitch, timber, and wool. Even in its best days, Puteoli could not balance what came into its harbour with what went out. What enabled Rome to buy far more than she sold was the great amount of capital which her conquests and the tribute levied on the provinces had brought her. Sometimes, indeed, the exportation of coin caused the government anxiety. Several Consuls tried to prevent the flight of silver and gold. The Gabinian Law of 67, which forbade provincials to go to Rome to get specie, may have had the same object. Cicero, in his consulship, ordered the custom-house officers at Puteoli to seize all gold and silver which anyone should try to take out of Italy.[3]

In Rome, whither most of this commercial movement converged, the chief business quarter was still, as we have seen, the Aventine Hill, with the immediate neighbourhood of the river-port. Retail trade was done in shops, which we may safely imagine to have been much like the shops at

[1] *Ad Att.*, i, 8-9.

[2] A. Merlin and L. Poinssot, in **XI**, vol. xvii, pp. 29 *ff.*; A. Merlin, *ibid.*, vol. xviii, pp. 5 *ff.*

[3] Cic., *Pro Flac.*, 67; *In Vatin.*, 12; *Ad Att.*, v, 21; *cf.* **LXIX**, pp. 252 *ff.*

Pompeii. The open places then existing, the Roman Forum itself, the Forum Boarium or ox-market, and the Forum Holitorium or vegetable-market, were doubtless used solely for daily and local trade. In Italy, Puteoli was at this time the principal centre of commerce. There the shipowners had their offices, docks, and warehouses, and there was a market on which the importer could offer his goods, almost before they were landed, either to the consumer direct or, perhaps, to a retailer.[1]

The commercial equipment of the Roman State was developed and perfected during the last centuries of the Republic. The Roman roads covered all Italy and were beginning to extend into some of the provinces. The Appian Way was continued beyond Capua, by Beneventum, to Brundisium. The Latin Way served the mountain country of the Hernici and Volsci, running parallel with the Appian Way as far as Capua. The Valerian Way crossed the Apennines and ended on the Adriatic at Aternum. The Flaminian Way ran to Ariminum and there joined up with the Æmilian Way, which went through the great plain south of the Po, by the colonies of Mutina, Parma, and Placentia. The Cassian Way ran north of Rome right through eastern Etruria. The Aurelian Way followed the coast of the Tyrrhenian Sea past Pisa into Liguria. Outside Italy, Rome had built the Domitian Way through south Gaul from the Alps to the Pyrenees, and the Egnatian Way from Dyrrhachium and Apollonia on the Adriatic (opposite Brindisi) right across the north of Greece by Epeiros and Thessaly to Thessalonica on the Ægean.[2] All over the East, in Asia Minor, in Syria, in Egypt, the old trade-routes, which had been so busy since the time of Alexander, continued to be used.[3]

But in the Mediterranean world the sea was the great road, for men and goods alike. By her decisive victory over Carthage, Rome became the mistress of the Western Mediterranean; nothing now stood in the way of her communications with Sicily, North Africa, the Iberian Peninsula, or Marseilles and the coast of Gaul. When she had triumphed over the

[1] **LXIX**, pp. 248 *ff.*, 255; Dubois, *op. cit.*, p. 268.
[2] **XVII**, *s.v.* " Via," pp. 783, 793 *ff.*
[3] See above, pp. 146 *ff.*

of the Empire. Now, it is fairly safe to say that the members of this oligarchy were far from being all of the purest Roman blood. That was not the least serious consequence of the Claudian Law.

IV

Movable Wealth. Capital

The movable wealth which became so important in Rome had other sources than commerce properly so called. The financial organization of the State and the fiscal organization of the provinces played an important part in its formation and development, to say nothing of the booty which victorious generals assigned to themselves, often in such quantities as to enable them to build up great fortunes.

Down to the end of the Republic, the administration of Roman finance was conducted on the system of giving out contracts or " farming " out branches of the revenue to the highest bidder. For the execution of public works, the distribution of military and other supplies, and transport by land or sea (all operations involving expenditure), and also for the working of State domains and the collection of taxes or rent (by which revenue was brought into the treasury), the government always preferred to employ middlemen rather than manage the business direct through its officials. Without dwelling on the advantages of these methods for the Republican administration,[1] we shall simply consider the very serious effects which this financial policy was bound to have, and had, on the economic affairs of Rome.

As the Roman State increased in size, as new provinces in the West and in the East were added to the Empire, public expenditure and revenue both rose enormously. More and more capital was needed if one wanted to bid at the auctions held by the Censors every five years. Rich as certain individuals had become by this time, their personal wealth was not sufficient for them to undertake such contracts. Collective resources were needed. These resources were obtained by the formation of financial societies or companies, which were joint-stock societies of " publicans." These companies were represented in dealings with the State by a

[1] LXVIII, p. 183.

responsible director (*manceps*). Their administration was conducted in Rome by a *magister* and in the provinces by men each of whom had the title of *pro magistro*. The subordinate staff was composed of collectors, couriers, accountants, and scribes, many of whom were slaves.[1] The shareholders of the companies of publicans were men of every class—Senators, who were forbidden by law to take personal part in the public auctions; Knights, who especially went in for big business of this kind; and even less wealthy citizens, who found these shares a profitable investment. It might, no doubt, happen that one company or another made no very great profit, or even that its balance sheet showed a deficit; but as a rule the opposite was the case. Sicily and Asia were scandalously squeezed in the interest of the publicans, who had such influence in Rome that it was impossible for the provincials to obtain justice, even in the most flagrant cases of organized, methodical spoliation.

Against these powerful companies, the governors of provinces dared not or would not lift a finger. Many of them preferred to follow their example. In fact, they were often driven to it by urgent personal necessity. To obtain the votes of the Comitia, they had had to incur huge debts, and the administration of a country like Sicily, Asia, Gallia Narbonensis, Further Spain, or Africa could bring them, provided they showed no moderation, honesty, or scruple, enough to enable them to pay off their debts and obtain considerable resources in the future. Very few resisted the temptation. Pompey, Crassus, Antony, and Cæsar, to name only the most famous, amassed great wealth in this way. Verres and Fonteius still live, thanks to Cicero, as typical specimens of the Proconsuls or Proprætors whose exactions were reported by their victims.

The most important economic consequences of this fiscal organization and the abuses to which it gave rise were the influx of much capital into Rome, the appearance of great fortunes, the most famous of which amounted to several hundred thousand pounds, and a movement of money hitherto unknown.[2] Certainly, it would be exaggerated and absurd to

[1] R. Cagnat, in **XVII**, *s.v.* "Publicani," p. 752.

[2] That of Crassus is roughly estimated by Frank at seven million dollars (£1,400,000).

called with banking operations, in which they often invested very large sums.[1]

The economic life of Rome at the end of the Republic was dominated by the progress and influence of movable wealth, of capital. This development was not so much due to a great rise of industry and trade strictly so called as to the triumphs of the foreign policy of Rome and the consequences of her victories. The origins of Roman capitalism must be sought in the huge spoils taken from the conquered peoples —booty seized during campaigns, tribute collected in the name of the victorious city, the unbridled, unscrupulous exactions of publicans and business agents, the profits of the financial operations of bankers, etc. Modern publicists and theorists make the capitalistic system responsible for the wars which turn the world upside down before our eyes. In the centuries immediately preceding the Christian era, it was war which gave birth to the capitalistic system in Rome, because the repeated victories of the legions led, at once or long afterwards, directly or indirectly, to the accumulation of immense wealth in the treasury of the State or the cash-boxes of individuals. In another form and in another direction, the creation of the Roman Empire affected the economic life of the time and contributed to its progress and development no less than Alexander's conquest of the East. At the end of the reign of Augustus, when the Empire was established and organized, the two great facts connected with the names of Alexander and Cæsar were, as it were, brought together and welded to one another. The Roman West and Alexander's East were now one single State; the unity of the Mediterranean was a fact. The economic life of the ancient world was entering upon the last phase of its development. The scattered, isolated efforts of the earlier periods would henceforward combine and fall into line one with another. From one end of the ancient world to another there would be a convergence of economic forces, which would make them far more effective and increase their product. The horizon of the activity and labour of man receded to the very boundaries of what the ancients called the Inhabited Earth, the Οἰκουμένη.

[1] R. Cagnat, in **XVII,** *s.v.* "Negotiator," pp. 45 *ff.*

PART IV

THE ECONOMIC LIFE OF THE ANCIENT WORLD UNDER THE ROMAN EMPIRE

CHAPTER I

NEW ECONOMIC CONDITIONS

THE establishment and organization of the Roman Empire united the ancient world, chiefly collected round the Mediterranean, for some hundreds of years. That political and administrative unity created new conditions for economic life, giving it a character and an amplitude which it had not had before. The peculiarity and importance of the period which covers the first centuries of the Christian era lie not so much in regional or local vicissitudes of economic life as in the coming together, interpenetration, and synthesis into one great organism of countries which had formerly been independent of one another. Their independence had, no doubt, not resulted in or been attended by complete isolation, but their power of production and capacity for consumption had not been exactly co-ordinated with those of the other parts of the world then known. Before we consider what the economic life of the ancient world was under the Roman Empire, therefore, we should determine what were the new conditions which it owed to the foundation of that Empire.

I

THE TERRITORIAL EXTENSION AND GENERAL ASPECT OF THE ROMAN EMPIRE[1]

At the end of the reign of Augustus, the Roman Empire, if one includes, together with the provinces properly so called, certain principalities or kingdoms united to Rome by ties of strict vassalage, embraced all the countries standing on

[1] **CV**, no. 12; **CVII**, no. xv.

the Mediterranean, in Europe, Asia, and Africa, from the end of the Euxine to the Strait of Gibraltar and from the coast of Spain and Gaul to the Nile Delta and the coast of Palestine. Attached to this great central mass were regions which faced west and north in Europe and south-east in Asia, and beyond these regions the Empire had already extended its boundaries to the Atlantic, the English Channel, the North Sea, the course of the Rhine, and part of the course of the Euphrates.

These frontiers were passed in the first and second centuries after Christ. The Empire was then augmented by the greater part of Britain, the Agri Decumates between the Rhine and the Danube, Dacia north of the lower Danube, part of Armenia, Assyria, Mesopotamia, and Arabia Petræa. It is true that some of these new provinces were soon given up, like Assyria, while others presently fell a prey to frequent invasions by the barbarians of Central Europe or the Parthians of Iran. The fact remains that the unity of the ancient world which Rome had brought about was not confined to the Mediterranean basin, but included countries washed by other seas, subject to other climates, and traversed by natural highways which did not approach the great Roman lake.

That unity, which had never been effected before, and has not been since the fall of the Roman power, did not wipe out the differences of every kind which existed between the provinces or homogeneous groups of provinces of the Empire. Those differences, geographical, racial, moral, and historical, were too marked and too deeply rooted in the soil or in the past for a foreign domination, however powerful, to be able to level them. Nor did Rome try to do so. From the peoples which she had subdued she demanded political, religious, and financial loyalty; she did not ask them to give up their traditions, to revolutionize their habits, to repudiate their sentiments, or to alter their activities, if those traditions, habits, sentiments, and activities did not conflict with the loyalty which she required. We find the Empire founded by Cæsar and Augustus as an aggregate of nations which preserved their own character, solidly welded together by an administration which was at once liberal and strict. In the first centuries of the Christian era, Italy was surrounded by a Roman Gaul, a Roman Spain, a Roman Africa, a Roman

Greece, a Roman Egypt; but in these double names each term kept its own value, and the special characters of the peoples did not disappear under a dull, uniform façade.

The result was that in this period economic life, within the actual bounds of the Empire, was at once national and international. If we look at each of the great regions of the Roman world in its historical domain, we do not by any means find that its economic ways were deeply or drastically altered. There was no violent break with the past. The same causes continued to take effect, the same methods of agriculture, industry, and trade determined the labour of men. The novelty lay in the increased intensity and larger output of that labour; in the attraction exercised on the various parts of the Empire, not only by Rome and Italy, but by the other parts; in the development of exchanges between North and South, East and West; and in the manner in which the Imperial policy started or guided the convergent movement of economic forces which had hitherto worked separately, sometimes combining but never obeying a single impulse.

This very important evolution was favoured, without any doubt, by the fact of the Empire itself and by the benefits which it shed for at least two centuries on all countries subject to Roman rule.

II

Peace Abroad and Security. Order at Home. Public Works

Not the least of these benefits was the establishment and maintenance of peace, peace on the frontiers opposite the barbarians and peace at home due to the extinction of civil war.

The military organization of the Empire, the establishment of strong frontiers, and the constant watching of the turbulent hordes which were always ready to invade peaceful and prosperous districts—all these measures ensured the security even of the more outlying provinces, so far as that was possible. The legions stationed at intervals along the Rhine, the permanent camps of Vindonissa (Windisch),

Augusta Rauracorum (Augst, near Basle), Argentoratum (Strasburg), Mogontiacum (Mainz), and Novæsium (Neuss, near Düsseldorf), and in the second century the Limes Germanicus with its many advanced posts, protected the whole of Gaul against inroad and pillaging by the Germanic tribes. The fortifications built in the north of Roman Britain, the Wall of Hadrian from the Solway to the mouth of the Tyne and that of Antoninus Pius connecting the Firths of Clyde and Forth, prevented the Picts and Scots of Caledonia from descending upon the countryside and handsome villas of the subject province. The fortified line of the Danube, guarded above the Iron Gates by the forts of Carnuntum (Petronell, east of Vienna), Aquincum (Budapest), and Singidunum (Belgrade) and covered below the gorge by Dacia, which Trajan conquered, and the great camp of Trœsmis (Iglitza on the lower Danube), protected the Balkan Peninsula from the Adriatic to the Black Sea. In the East, Syria and the neighbouring country, which were exposed to attempts at conquest and mere raiding by the Parthians, enjoyed comparative security, thanks to the nature of the back-country, the defence-works established along the Euphrates, and the subsequent creation of the provinces of Mesopotamia and Arabia. In North Africa, the cities of Tripolitana, the oases of southern Tunisia, the Numidian plateaus, and the Mauretanian Tell, having been liberated from the danger of sudden raids by the victories of Rome over the Garamantes, the Numidian chief Tacfarinas, the Gætulians of Numidia, and the Moors of Tingitana, experienced a hitherto unknown prosperity for over two hundred years under the protection of the legion at Lambese and the numerous detachments which held well-chosen strategic points far out in the desert or just on the edge of the cultivated land. So, under the successors of Augustus and down to the anarchic times of the third and fourth centuries, legions, auxiliary troops, forts, and outposts formed a strong, continuous barrier, behind which the economic activity of the world could go on in all confidence.

Inside the Empire, political unity, the establishment of order, the action of a sometimes strict authority, had an equally efficacious and beneficent influence on economic life. No doubt, the Roman world still experienced some shocks

—in 68-69, the terrible year which followed the death of Nero; from 193 to 197, during the wars of Septimius Severus, Pescennius Niger, and Clodius Albinus; and still more frequently in the third century. No doubt, too, local rebellions bathed one or another part of the Empire in blood—those of the Jews from 66 to 70 and from 131 to 135, that of the Batavians and several Gallic tribes of Belgium in 69-70, and that of the Britons and Queen Boudicca or Boadicea in 61. No doubt, lastly, in the third and fourth centuries, the military anarchy, the attempt of Postumus and Tetricus to create a Gallic Empire, the revolt of Palmyra under Odenathus and Zenobia, and the rivalries of pretenders and civil wars which preceded the final accession of Constantine from 306 to 323, were all upheavals which disturbed the public peace. It is true that one cannot regard the first four centuries of our era as an age of perfect and uninterrupted tranquillity for all the provinces of the Empire. Nevertheless, in East and West alike, this period, or at least the first two centuries, seems remarkably peaceful, if we think of the constant wars of the Greek cities and the Hellenistic rulers, of that of Mithradates with the Romans, of the civil wars which ended in the victories of Cæsar and Augustus and were followed by violent spoliation, or of the brigandage and piracy which used to go on by land and sea.

Moreover, the governors of the provinces, Proconsuls, Legates, or Procurators, being less free in their action, more closely supervised, and compelled to render account of their administration, not to an Assembly or law-courts in which they found accomplices more often than impartial judges, but to a chief who was often severe and sometimes cruel, ceased to squeeze the provinces as Verres and Fonteius had done. Without disappearing altogether, their exactions became less frequent. In addition, the yearly meetings of the Provincial Assemblies gave the delegates of cities an opportunity to express their opinion of the conduct of governors. Consequently there was less disturbance and violence, more order, and more honesty in the administration; such were the advantages conferred on the provinces by the organization of the Empire, and from one end of the Roman world to the other these advantages increased the return of labour in all its forms.

The Roman government itself directly encouraged economic progress by the impulse which it gave to public works. The working of mines and quarries, the construction of numerous roads and bridges, and the building, restoration, or improvement of harbours, often big ones, supplied industry and trade with raw materials and an equipment which they had hitherto lacked. The spread of Roman law, which safeguarded the economic dealings of the peoples of the Empire as it did their civil relations, also had an influence on production and on all the various transactions by which natural commodities and manufactured goods are brought from the producer to the consumer.

The new conditions, geographical, political, administrative, and social, which the constitution of the Roman Empire ensured for economic life were not bought at too high a price, in the shape of new burdens. For at least two centuries after the death of Augustus, the ancient world was undeniably prosperous, except in special cases, such as that of Greece, which was deserted and almost depopulated, or Palestine, which was ruined by bloody insurrections and still bloodier repressions. There was compensation for such wastage in the extension of productive land in many parts of the West, in the rapid development of industry there, and in the unprecedented increase of trade. Numbers of new cities sprang up in North Africa, in the Peninsula, in Gaul, and along the Rhine and Danube, as they had done in Asia Minor and Syria after the conquest of the East by Alexander. The middle classes of the *municipia*, who owed their wealth to landed property, industrial undertakings, and trade, gave evidence of their prosperity by everywhere giving large sums for the embellishment of towns, charitable works, and foundations, which were doubtless inspired by vanity more often than by a sincere feeling of generosity or gratitude. Outside the towns, the surviving remains of country-houses and farm buildings show what a place rural life had in the activity of the provincials. There are countries round the Mediterranean which for fifteen centuries have not recovered the economic activity of which they were then the scene.

III

The Influence of Ancient Economic Life outside the Empire

The effects of the new economic conditions were not confined to the regions composing the Roman Empire. They were felt far beyond the frontiers. The Mediterranean world at that time had frequent and regular commercial dealings with Asia, the Far East, and Central and East Africa.

The geographical knowledge of the Greeks and Romans extended a long way towards the Far East and the Equator. To see the progress accomplished, one has only to compare a map of the world based on what Strabo knew at the beginning of our era with one based on Ptolemy's description of the world five hundred years later. Strabo does not go further east than the Ganges basin in India and the island of Taprobane (Ceylon), or further south-eastwards than Spice Cape (Guardafui), and he has no definite ideas about the interior of Africa. Ptolemy's account embraces Indo-China (the Golden Chersonese of the ancients) and reaches the southern seaboard of China, while in Africa he knows (vaguely no doubt, but still he knows) of the lakes of the country about the sources of the Nile, and the Niger further west. Other sources of information corroborate the conclusions suggested by the examination of these two maps. "At Kananor" (on the coast of Mysore) "numerous Roman gold coins of the Julio-Claudian epoch have been found, formerly exchanged against the spices destined for the Roman kitchens. On the island Salice, the Taprobane of the older Greek navigators, the modern Ceylon, in the time of Claudius a Roman official, who had been driven thither from the Arabian coast by storms, had met with a friendly reception from the ruler of the country, and the latter, astonished, as the report says, at the uniform weight of the Roman pieces of money in spite of the diversity of the emperors' heads, had sent along with the shipwrecked man envoys to his Roman colleague. . . .

"It was not till later apparently that navigation was extended as far as that large and productive island, in which on several occasions Roman coins have come to light. But coins are found only by way of exception beyond Cape

Comorin and Ceylon, and hardly has even the coast of Coromandel and the mouth of the Ganges, to say nothing of the Further Indian peninsula and China, maintained regular commercial intercourse with the Occidentals. . . . By sea, certainly, individual navigators reached accidentally or by way of exploration at least to the east coast of Further India and perhaps still farther; the port of Cattigara known to the Romans at the beginning of the second century A.D. was one of the Chinese coast-towns, perhaps Hang-chow-foo at the mouth of the Yang-tse-kiang. The report of the Chinese annals that in 166 A.D. an embassy of the emporer Antun of Ta- (that is Great) Tsin (Rome) landed in Ji-nan (Tonkin), and thence by the land-route arrived at the capital Lo-yang (or Ho-nan-foo on the middle Hoang-ho) to the emperor Hwan-ti, may warrantably be referred to Rome and to Marcus Antoninus. This event, however, and what the Chinese authorities mention as to a similar appearance of the Romans in their country in the course of the third century, can hardly be understood of public missions . . . but possibly individual captains may have passed with the Chinese court as messengers of their government."[1]

In Africa, expeditions starting from the Mediterranean coast, chiefly from Leptis Magna, crossed the Sahara and made their way as far as the Sudan. Certainly, as M. Gsell so justly remarks, "these relations between the Syrtes and the Sudan did not advance geographical knowledge."[2] There are many confusions and inaccuracies of detail in Ptolemy's account of the interior of Libya.

But it is not inaccurate and uninteresting details that Marinos of Tyre gives us about the journeys of Septimius Flaccus and Julius Maternus, who were doubtless contemporaries of Domitian. The former travelled southwards for three months, beyond the country of the Garamantes, that is, the Fezzan. The latter came, after four months, to the country of Agisymba, that is, the oasis of Asben, in the part of the Sahara near Lake Chad, and reached the country of the Ethiopians, the Negroes, where the rhinoceros is to be met.[3] It seems clear that the Roman explorers reached

[1] **XXXIV**, *Prov.*, English vol. ii, pp. 301-302.
[2] *La Tripolitaine et le Sahara au IIIe siècle de notre ère*, p. 6.
[3] Ptol., i, 8, 4; **LXXXIV**, 2nd ed., p. 42.

Equatorial Africa. Further east, two centurions with a detachment of Prætorians went up the valley of the Nile as far as the Bahr el-Ghazal.[1]

In the north of Europe, Roman trade has left traces on the shores of the Baltic, in the Scandinavian peninsula, and in Livonia. These are the furthest points which it touched.[2]

So the economic field of action of the ancient world was widened in many directions; only the Atlantic opposed an impassable barrier.

The establishment of the Roman Empire, then, did not affect only the countries within its own borders. It caused both the political and the economic horizon of the classical world, of Græco-Roman civilization, to recede a long way, eastward, southward, and northward. A similar but much smaller progress had been brought about by the expedition of Alexander, an offensive movement of Greece towards the East. At the beginning of our era, the initial cause of the evolution was a work of internal organization which gave economic forces a more vigorous impulse, more regular working, and a greater power of radiation.

It is in the second century of the Empire, under the government of the Antonines, that one must take one's stand in order to observe this expansion of ancient economic life; it was then that the new conditions, deriving from the very organization of the Roman Empire, were most completely realized and took full effect.

[1] Sen., *Quaest. Nat.*, vi, 8; *cf.* Pliny, *N.H.*, vi, 35.
[2] **XCVII**, p. 304.

CHAPTER II

AGRICULTURE AND STOCK-BREEDING. PROPERTY AND AGRICULTURAL LABOUR IN THE ROMAN EMPIRE

By collecting so many countries which had once been independent and separate states under her political dominion, Rome had not altered the natural laws to which they were subject. By giving all these countries a single, though elastic, administrative organization, she had not changed the nature of the soil or the conformation of the land or the climate. After the victorious city had done her political and administrative work, agriculture, stock-breeding, hunting, and fishing, in other words the flora and fauna, were determined by the same old geographical conditions as before. The crops grown and the animals reared continued to be, for the most part, those characteristic of the Mediterranean region, with those which belonged rather to the Atlantic type in Western and Central Europe and those related to the tropical or desert type in the east and south.

Does this mean that, in the domain of agriculture and stock-breeding, there was no economic evolution, and that all that we have to do is to point out briefly the continuity, the permanence, of previous conditions, as they had existed in Greece and the East in Hellenistic times, in North Africa under the supremacy of Carthage, in Italy, Spain, and Gallia Narbonensis before the establishment of the Empire, and in the north and centre of Europe during the Bronze and Iron Ages down to the Roman conquest? Not at all. The agriculture and stock-breeding of the ancient world were transformed in the first centuries after Christ, and these changes were without exception due to the action of man.

The relative importance of corn and fruit growing and stock-breeding fluctuated in many districts, here in favour of cultivation and there in favour of stock-breeding; the area devoted to one crop or another increased or diminished; the result was a new distribution of the centres of agricultural production. It is probable that in several provinces the forests were considerably reduced. All these changes were directly or indirectly the work of man, some being intended, prepared, and carried through by him and others being the consequences, perhaps not very clearly foreseen, of military conquests, territorial annexations, or political reforms.

Conquests, annexations, and reforms greatly increased the economic field of action. Vast areas, formerly uncultivated or neglected by semi-barbarous populations, began to be made productive. In these districts, as in countries which had been cultivated before, methods of working were improved; more detailed study and intelligent utilization of the soil and climate made it possible to increase the return.

Lastly, under the influence of the institutions and law of Rome, the legal and economic organization of property and the practical organization of agricultural and pastoral labour were perfected. Local diversities in this organization were reduced. The rural economy of the ancients came in the end, after long centuries of divergent customs, to a relative unity in conception and application.

I

Crops, Livestock, Forests, Hunting, and Fishing

The Roman world did not discover any new crops. Corn, chiefly wheat and barley, the vine, the olive, vegetables round the towns, fruit-trees more or less everywhere, flax here and there, meadows and fodder-plants in damp ground, and flowers and certain luxury-plants in the gardens continued to be grown by farmers and landowners. The papyrus and lotus in Egypt, silphium in the region of Cyrenaica, and the lotus in Tripolitana kept their local importance.

So, too, stock-breeding remained unchanged in its principal forms. Oxen, sheep, and pigs still held the first place, with horses, mules, and asses. Bees and their hives, the many breeds of poultry and their yards, even fish and their ponds were the object of particular care. For some centuries elephants had appeared on the battlefield, not only in the East, but even in Italy, where Pyrrhos had used them against the legions at the beginning of the third century B.C. Under the Empire, Rome did not use elephants in warfare, but they were harnessed to triumphal cars, and above all they appeared in the games of the circus and the combats of the amphitheatre. We cannot speak of elephant-breeding in the strict sense. These animals were chiefly captured wild in Asia and in parts of Africa; but there were depôts where they were trained near Rome.[1]

Camels were used by the Romans for public transport, both in war and in peace, and they also drew cars in the circus.[2] The Romans imported them from Arabia and the plateau of Iran, and some may have been bred in Egypt. The camel does not seem to have become acclimatized in Africa before the end of the Empire. Ammianus Marcellinus relates that in 363 Romanus, the Count (*i.e.*, military governor) of Africa, having been begged by the inhabitants of Leptis Magna, a big town in Tripolitana, to protect them against the inroads of desert raiders, ordered them to supply 4,000 camels for the expedition. " The people of Leptis," adds Monsieur Gsell, " did, it is true, declare that they could not supply this number; but they had just been very thoroughly raided. In ordinary times they could no doubt have easily satisfied the demand of Romanus if they had wished."[3] The episode shows that camel-breeding was highly developed in Roman Africa, at least in the parts near the Sahara.

Elephants and camels—these are the only novelties which we can find in the stock-breeding of Imperial times. Moreover, these animals were very little used, and the employment of them seems to have been of no economic importance, or, indeed, of an economic character at all.

There being no great innovations in the crops and animals raised, what strikes one when he studies the agriculture of the Roman world in the first centuries of our era is the development which took place in the diffusion and relative distribution of the chief methods of farming and the most widely grown crops. On the one hand, in certain provinces, Africa, for instance, land which had once been entirely left to livestock and nomadic tribes became covered with crops;

[1] S. Reinach, in **XVII**, *s.v.* " Elephas," p. 543.
[2] E. Saglio, in **XVII**, *s.v.* " Camelus," pp. 856 *ff.*
[3] **LXXI**, vol. i, pp. 59 *ff.*; Gsell, *La Tripolitaine et le Sahara au IIIe siècle de notre ère*, p. 10.

on the other, in Italy and Greece vast tracts which had once been cultivated were transformed by the misfortunes of the times or by the neglect of the great landlords into pasture, when they were not utterly abandoned. The production of corn, especially of wheat, went down, whereas the vine and olive gained very much ground in the East and West. The result was that there were famines, which led to grave disorders, chiefly in the Greek and Asiatic provinces. The Imperial government could not remain indifferent to these economic happenings. Measures were taken to encourage corn-growing and to stem the advance of wine-growing.[1] Furthermore, the exportation of wine had become one of the principal resources of Italy, but in this domain the competition of the provinces became more and more dangerous, and almost disastrous. Domitian tried to remedy both this danger and the consequences of the disfavour into which corn-growing had fallen. Suetonius relates that he issued an edict forbidding the planting of any more vines in Italy and ordering that only half of the vineyards then maintained in the provinces should be allowed to survive. The provincials made such an outcry that the Emperor had to repeal the edict, or at least to restrict the application of it.[2] Olive-growing, on the other hand, was fostered in many regions, in Dalmatia, in Spain, in Africa.

The works written under the Empire dealing wholly or in part with agriculture and stock-breeding furnish some kind of evidence, by their character, regarding this development. Of the thirteen books of Columella's *De Re Rustica*, not one is devoted from beginning to end to corn-growing. Wine-growing, on the other hand, takes up two and a half books, and fruit-trees, including the olive, have a big place, the end of Book III and a great part of Book VI. Livestock in its various forms—large and small cattle, poultry, and bees—occupies four books (VII to X). Book XI, which is written in verse, deals entirely with gardens. In the encyclopædia of Pliny the Elder, corn is treated in detail only at the beginning of Book XVIII, while Books XII to XVII are full of information about trees and tree-growing of every kind. The

[1] **XCVIII**, pp. 188 *ff*.

[2] *Ibid.*, p. 189; *cf.* Gsell, *Essai sur le règne de l'empereur Domitien*, p. 153.

whole of Book XIV is taken up with the vine, and Book XV with olives and fruit-trees. In the fourth century, Palladius Rutilius Taurus Æmilianus, the author of a treatise *De Re Rustica,* in which the various labours of the farm are described month by month, observes the same proportion. At intervals he mentions corn-growing—wheat, spelt, barley—and dwells at much greater length on the vine, the olive, fruit-trees, gardens, meadows, and livestock.

There is no doubt, therefore, that the Roman Empire went through what may be called a corn shortage. The shortage was further aggravated by the policy adopted by the Emperors. For the sake of their popularity and indeed of their safety, their chief concern was to keep the capital fed. Accordingly, they forbade the corn of Egypt to be exported anywhere but to Rome, and reserved for the Roman populace the harvests which were gathered on the immense Imperial estates of Sicily, Africa, Spain, and Gaul. The quarter-master's branch of the army appropriated all or nearly all of the output of southern Russia for the feeding of the armies of the East. Local and regional production had to suffice for the feeding of other cities and agglomerations.[1]

By the side of the cultivated land and pasture, forests took up considerable areas. The conquest of Western and Central Europe put the Romans in possession of well-wooded regions, Gaul, Noricum, Rætia, and in general all the mountainous districts from the Pillars of Hercules to the valley of the Rhine and from the tip of Brittany to the mouths of the Danube. The African Tell was likewise covered with dense woodland. In addition to high forest, there was plenty of coppice on the great estates of the Emperors and wealthy individuals, in the *saltus* of Italy and the provinces, the organization of which has been revealed by numbers of inscriptions. Wood for house-building, shipbuilding, joinery, coach-building, and cabinet-making was furnished by these vast forests, both to the Imperial government for the requirements of the armies, fleets, and public works and to private industry.[2] Characteristic evidence of the importance of the forests and their place in the life of the peoples of the Roman Empire is supplied by the innumerable hunting-scenes

[1] **XCVIII**, p. 188.
[2] A. Jacob, in **XVII**, *s.v.* "Materia," "Materies," pp. 1626 *ff.*

sculptured on funerary monuments and represented in mosaics.[1] Hunting was practised either as a sport or to capture the beasts needed for the bloody games of the circus and amphitheatre.

River-fishing and, above all, sea-fishing were practised actively in the Roman world. The big rivers, Po, Rhine, and Danube, and the lakes of northern Italy sent their fish to Rome. In the Atlantic, whales and cachalots, sharks, and seals were captured, sometimes at great risk. On the Mediterranean, the tunny-fisheries were perhaps the most profitable and best organized. The countries on the coast of the Euxine, which sent large cargoes of salt fish to Greece, were reinforced in the Roman period by southern Spain, where many ports lived by the salt-fish industry, and the same business had also developed at certain points on the coasts of Sardinia, Sicily, and Tripolitana.[2] M. Lafaye has recently shown that in the Roman period, on the coasts of Italy and Roman Gaul, fish-ponds, which were traps for fish rather than mere preserves, were constructed.[3] So marine fauna, like terrestrial fauna, had its place in the economic life of the Roman world.

II

Agricultural Progress

While there was little change in the sort of crops grown and beasts raised, we should note the progress made in the Roman period by agriculture in every direction and in many countries.

Agriculture gained new ground from the forests, the swamps, and the desert. We have not sufficiently detailed and local evidence to estimate the amount of deforestation done at that time. If it is too much to say, as has too often been done in the case of Gaul, that whole forests disappeared, it is at least certain that enormous holes were made in them. What M. Camille Jullian says of Gaul may, I think, be applied to many of the western provinces. " New villas and sanctuaries were built on the fringes of the woods, or even by the side of a spring in a clearing inside, and a building of that kind,

[1] G. Lafaye, *ibid.*, *s.v.* " Venatio," pp. 696 *ff.*
[2] G. Lafaye, *ibid.*, " Piscatio " and " Piscatus," pp. 491 *ff.*
[3] In *Bull. de la Soc. des Antiquaires*, 1919, pp. 183 *ff.*; 1921, pp. 293 *ff.*

which constantly extends its offices, its parks, its meadows, its orchards, and its kitchen-gardens, is a cause of ruin to the forest round it."[1] Because later, after the invasions of the fifth century, the forest regained ground over the ruins left by the barbarian hordes, we must not conclude that it had always occupied that ground and had never been expelled from it. In Britain, in the valleys of the Peninsula, where there were so many thriving cities and townships, in Rætia, in Noricum, in Pannonia, in many parts of Illyricum, where the population now for the first time became acquainted with city life, and on the slopes of the African Tell, high forest and coppice had to open out and recede to make room for plough-land, orchards, farms, and country-houses.

The swamps round the towns and in the country were attacked. Drainage-works which carried off the stagnant water made it possible to give them a drier, solider, healthier soil. This work of sanitation was sometimes given to veterans. At the beginning of the reign of Tiberius, one of the ringleaders of the mutiny of the legions of Pannonia roused the soldiers by telling them that the veterans were being given "fields" in the shape of damp marshland and uncultivated hillsides, and were being sent into all kinds of countries—*trahi adhuc diversas in terras, ubi per nomen agrorum uligines paludum vel inculta montium accipiant.*[2] With reference to Gaul, M. Jullian is inclined to believe that the marshes of the Saintonge, Vendée, and Flanders were partly drained round farms whose centre was on a more substantial island rising above the low ground.[3]

Lastly, the desert—or rather the land which had hitherto been left untilled because of the climate, conformation, or lack of human labour—even the desert gave place to rich estates scientifically worked. It is true that in certain districts which had once been prosperous and fertile, such as central and southern Italy and most of Greece, much land was abandoned and reverted to waste; but the newly conquered or newly occupied provinces offered the farmer vast areas, previously neglected by man, which were covered with rich harvests for several hundred years. We can best observe the change in North Africa and in Europe.

[1] XCVI, vol. v, p. 179.
[2] Tac., *Ann.*, i, 17.
[3] XCVI, vol. v, p. 177.

In North Africa, at the beginning of the Christian era, nomadism prevailed except on the coast and in the seaward valleys of the Tell. Strabo declares that in the interior there was nothing but a succession of mountains and deserts, and that the life of the peoples between Gætulia and the Mediterranean coast was like that of the nomadic Arabs.[1] Pomponius Mela, a contemporary of the Emperor Claudius, speaks of barren, uninhabited tracts in Numidia.[2] After enumerating the cities lining the coast from the Pillars of Hercules to Cyrenaica, he goes on:

"The peoples who live on the coast have adopted our manners and ways in everything, except that a few have preserved their old language and the gods and rites of their ancestors. Those who live immediately inland from them have no cities, their dwellings are called *mapalia*, and their life is hard and rude. . . . Further inland, there are still ruder men, who follow their herds as they wander about, go from pasture to pasture carrying their shelters with them, and spend the night where they happen to be when the day ends."[3]

A century later, these same regions, which Strabo and Pomponius Mela described as the domain of pastoral nomads, were covered with rich cultivated land and flourishing cities. According to the nature of the soil, the climate, and the water-supply, corn, vines, olives, and fruit had taken possession of the ground. Round the many cities, agricultural life had taken the place of pastoral life. The magnificence of such ruins as those of Sufetula, Thelepte, and Ammædara in Tunisia and of Madaura, Theveste, Timgad, and Lambese in Algeria still bears witness to the wealth which developed in these regions where hitherto nothing had lived but herds of beasts and their miserable owners. Immense olive-groves covered the plateaus, and between the towns the country was covered with farms and their oil-presses.

We have less abundant and less detailed information about the progress of agriculture in most of the European provinces. It cannot, however, be denied that agriculture advanced greatly in some parts of Spain, for example the centre and north-west, where city life took the place of the old tribal organization, at least to some extent; in Britain, where many great agricultural estates, farm-houses, and

[1] xvii, 3, 9 and 19.
[2] i, 6.
[3] i, 8; for this aspect of Africa at the beginning of the Empire, *cf.* J. Toutain, in **XXII**, pp. 319 *ff.*, esp. pp. 323-25.

wealthy villas sprang up round the towns; in Belgium, in the Agri Decumates, in Noricum, in the inland valleys of Dalmatia, and all along the Danube from Pannonia to the shores of the Euxine.[1] Here the influence of the legions and the work of the veterans seem to have played the chief part. Nor, without doubt, was it by mere chance that the ringleader of the mutiny in Pannonia at the beginning of Tiberius's reign reproached the government with giving the veterans the barren sides of mountains for fields. These regions of Western and Central Europe may not have undergone a change to be compared to that which North Africa presents. But agriculture certainly made great progress there, either at the expense of stock-breeding or by the reclaiming of hitherto unexploited land.

The quality of work, the return, the methods of farming were improved. On observing the contrast which existed recently and still exists in certain Mediterranean districts between the present poverty of agriculture and the undeniable evidences of ancient wealth, some thought at first that natural conditions, particularly the climate, had changed since the beginning of our era; that the rainfall had become less frequent and plentiful. This theory was held for a long time in respect of Africa in particular, and one might be tempted to apply it to many parts of Syria. But further study, inspired by many archæological discoveries and a careful reading of the ancient authors, has shown that this economic decline has everywhere been the doing of man, as was the prosperity of earlier times and as the revival of a wealth which has been eclipsed for many centuries is today and will be more and more in the future.

The climate has not changed perceptibly in any part of the Mediterranean basin. The deforestation from which many mountain ranges have suffered may have accentuated the torrential character of many watercourses, but it does not seem to have had any effect on the frequency, distribution, and character of the rainfall, which chiefly depends on the direction of the wind. What did change, after the fall of the Roman Empire, was the work of man. The economic ruin which long lay on these once flourishing countries was

[1] **XCVIII**, p. 296.

due to the abandonment of the technical methods and processes which were formerly employed.

We cannot say whether the object was attained as a result of co-ordinated research and experiment or by a series of practical shots, but there is no doubt that farmers succeeded, in most provinces, in developing the crops which best suited the natural conditions. It is true that in Gaul wine-growing and olive-growing were subjected to many distractions by Roman legislation, being forbidden towards the end of the Republic in the interest of Italian landowners, authorized, at least in practice, by Cæsar and Augustus, and restricted again (the vine, at least) from Domitian to Probus; but these two occupations, which suited the climate of the country, first developed under Roman rule, and chiefly under the Empire,[1] Gaul having previously been almost entirely a corn-growing and pastoral country.[2]

Perhaps the most characteristic example of the application of intelligent method to agriculture was the incredible development of olive-growing in semi-desert countries, such as the high plateaus of Algeria and Tunisia, the east face of the chain of Lebanon, and the country stretching from Hamath and Homs to Palmyra. The irregularity of the rainfall and the superficial character of the soil, which was chiefly composed of sand and was very pervious, did not allow of the cultivation of plants with short roots, and of corn in particular. On the other hand, since the moisture remained in the lower strata, at a depth varying between about half a yard and a yard below the surface, the growing of trees, whose roots went down far enough, was certain to prosper. The experiment has been made in the gardens round Sfax in Tunisia. "The same soil," M. Paul Bourde writes, "remains barren or is covered with vigorous vegetation with plentiful fruit, according as one sows it with cereals, whose roots, not going below the surface soil, perish in the dryness, or plants it with trees whose roots run deep into the ground. All fruit-bearing species which like dry climates succeed in these gardens and would succeed in the rest of central Tunisia, since the climate and soil there are similar."[3] These conditions, which

[1] **XCVI**, vol. v, pp. 183 *ff.* [2] *Ibid.*, vol. i, pp. 84 *ff.*

[3] *Rapport sur les cultures fruitières et en particulier sur la culture de l'olivier en Tunisie*, pp. 16 *ff.*

are peculiar to a vast district in North Africa, were observed and turned to account under the Roman Empire. It was by the olive and thanks to the olive that the Byzacene district, described by Sallust as a desert and barren country at the time of the Jugurthine War, became thriving and populous in the first centuries of our era. We can now understand the Arabian historians when they relate that when the first Mussulman conquerors arrived in North Africa one could ride in the shade through an uninterrupted line of villages from Tripoli to Tangiers.[1] That shade cannot have been given by high forest, since an uninterrupted line of villages is mentioned. What there was, was an immense olive-grove.[2]

The prosperity of olive-growing is also attested by numerous and conclusive indications east of the Lebanon and the Orontes. " While at the present day olives scantily grow only in the valleys of the Lebanon abounding in springs, the olive woods must formerly have stretched far beyond the valley of the Orontes. The traveller now from Hemesa to Palmyra carries water with him on the back of camels, and all this part of the route is covered with the remains of former villas and hamlets. . . . The Austrian engineer, Joseph Tschernik, found basalt-slabs of oil-presses not merely on the desert plateau at Kala'at el-Hossn between Hemesa and the sea, but also to the number of more than twenty eastward from Hemesa at el-Ferklûs, where the basalt itself does not occur, as well as numerous walled terraces and mounds of ruins at the same place; with terracings on the whole stretch of seventy miles between Hemesa and Palmyra."[3]

It is not a mere chance that we find, in the regulations for the farming of various great African estates, that privileges are granted to farmers who shall plant or renew vineyards, olive-groves, and fig-orchards. The encouragement thus given to tree-growing is explained both by the natural conditions of soil and climate, which were favourable to this form of farming, and by the wish to lessen the danger of a sudden, violent flow of surface-water, which was often disastrous.[4]

1 Tissot, *Géographie comparée de la province romaine d'Afrique*, vol. i, p. 278.
2 *Cf.* J. Toutain, *Les Cités romaines de la Tunisie*, pp. 40 *ff.*
3 **XXXIV**, *Prov.*, English vol. ii, p. 136 and n.
4 J. Toutain, in **XVI**, 1899, pp. 311 *ff.*

I have dwelt on the very suggestive case of olive-growing, because we have quite explicit evidence of its importance in many provinces of the Empire. One may presume that this case was not unique, that other forms of farming were practised, encouraged, and developed on the same lines, and that in this way vast areas, unsuited to the growing of corn, vegetables, or grass, were none the less made productive.

Another advance in agriculture was made under the Empire by the methodical and judicious use of the water supplied by nature. The part played by irrigation in the fertility of certain districts and the prosperity of certain crops may, perhaps, have been exaggerated. This method of utilizing the water does not seem to have been applied to large areas in any province of the Roman Empire, except perhaps Egypt. No vestige or evidence has been found of any such works in North Africa, the Peninsula, the rich valleys of Asia Minor, or Syria. Wherever it has been possible to follow up one of the many aqueducts or water-conduits constructed in the first centuries of the Empire, anywhere in the Roman world, it has been found to end either in a city or village or in a villa or farm. The water was therefore intended for the use of human beings and livestock. Irrigation was only applied to the growing of vegetables, flowers, and certain trees, and perhaps also to meadows and fodder-plants, round cities and on farms. Pliny has given us a valuable piece of information about the water-system of the African oases. The spring which now supplies Gabes was already the source of the wealth of Tacape in ancient times, its water being distributed among the towns-people at given times of the day—*certis horarum spatiis dispensatur inter incolas.*[1] A celebrated inscription, found at Lamasba, west of Batna, tells us that it was the same on the plateaus of the Constantine district. Here again, irrigation was practised in the immediate outskirts of a city; the actual wording of the regulation allows one to suppose that it supplied, not cornfields, but gardens, orchards, or meadows.[2]

One must not, therefore, exaggerate the importance of irrigation in agricultural practice so as to give a false impression. This does not mean that the work done by the Romans in this respect did not have a great and beneficial effect on

[1] *N.H.*, xviii, 22 (51). [2] V, vol. viii, 18587.

rural economy. By supplying human beings and domestic animals with the drinking-water which they needed, it made it possible to farm regions in which there could have been no settled life or farming without it. Without the tappings of springs, the dams across the rivers, and the huge, solid cisterns and reservoirs, which ensured a sufficient supply of water in every season, the highlands of Tunisia and Algeria and the half-desert fringes of Syria would have remained uninhabitable and would never have known the prosperity of which so many vestiges remain. Irrigation affected agriculture in much the same way as the progress of city life did; it contributed indirectly, but effectively, to extending the areas which could be farmed.

III

Property and the Organization of Agricultural Labour

One result of the combination of all the territories conquered and annexed by Rome in a single state was that agricultural property, from one end of the ancient world to the other, received, not indeed a uniform constitution, but characteristics and methods whose diversity was, as it were, systematized. That diversity, with all its delicate gradations, was based on the distinction between the territory of the Roman City and that of the cities, kingdoms, and tribes outside the Roman City, which had been conquered by her or had become her allies. It had important economic and legal consequences.

The Roman soil, *ager Romanus*, at the beginning comprised only the City's own territory. In the eyes of the Roman law, it alone could be held in complete ownership (*ex jure Quiritium*), and it was not liable to any tax. In the last century of the Republic (we cannot tell the exact date or circumstances), the whole of the soil of Italy was assimilated to the original *ager Romanus*, the notion of *ager Italicus* took the place of that of *ager Romanus*, and the term *jus Italicum* designated the legal condition of land in the provinces which was given the privilege of full ownership, the occupiers of which had *dominium ex jure Quiritium*.

Outside Italy, the ground was, on principle and save for exceptions, of provincial status, that is, it was subject to a tax and did not carry full ownership with it.[1]

But we should be giving much too simple a notion of the organization of agricultural property in the Roman Empire if we confined ourselves to pointing out this fundamental distinction. There were different categories of land in the provinces. In every region, the Roman State under the Republic and the Imperial government after Augustus had reserved for itself the direct exploitation of certain domains, usually very large, some of which came from the property of kings to whose rights Rome had succeeded, others from the territory of cities which had fought for their independence to the end and had consequently been deprived, to some extent, of their official existence, and others from arbitrary confiscations effected by the Emperors, as was done by Nero in Africa. The greater part of the land in the provinces, in theory the property of the victorious city, had been treated in one of three ways. (i) Some had been allotted to colonists, who were either lumped together so as to form an urban agglomeration or provided individually (*viritim*). In either case, the conceded land had been officially measured and entered in the survey-register. (ii) Some had been sold by the Quæstors to private individuals. This land usually took the form of very large estates. We know that there were properties of this kind in Africa round which one could not ride a horse in a day. They were independent of the municipal administration, and their owners often went to law with it. (iii) Some was left to its old owners, who had submitted to Rome quickly and had been allowed to keep their cities, villages, or tribal organization. This was the most usual fate of land in the provinces.

We must also give a special place to the territory of cities which, by a somewhat rare favour, Rome treated as allies and not as subjects (*civitates fœderatæ*). This territory was regarded as being outside the Empire and enjoyed an independence which, however, was more nominal than real.

The legal relationship of the possessors of the soil to these various categories of land was not the same as that of the possessors of Roman or Italian land. Obviously, the State

[1] **XCI**, pp. 250 *ff.*

or the Emperor had full and complete ownership of public land, whether it was assigned to the *fiscus* (that is, to the State) or to the private patrimony of the Emperor, and the owners of these domains did not pay any tax on them. But land in the provinces, whether occupied by Roman citizens or by *peregrini*, whether comprised in the territory of a city or not, was not susceptible of full ownership, unless *jus Italicum* was granted. It had to pay a land-tax, unless the Imperial government granted it the privilege of *immunitas*. Even the land of the colonies, although distributed to veterans who had been given the Roman citizenship or to citizens hailing from Italy, was subject to the same rule; the Gromatici speak of *agri colonici stipendiarii* by the side of *agri colonici immunes* and *agri colonici juris Italici*. The rule was that all land in the provinces, even when possessed by a Roman citizen, had to pay land-tax. The real meaning of the tax was to recall and express the eminent ownership of the State over conquered lands. Exemption from that tax, like the concession of ownership *ex jure Italico* or *Quiritium* over provincial land, was an exceptional privilege.

The only land in the provinces to be legally free from this condition was that of the allied cities. It did not pay land-tax to the Roman State, and its possessors had full ownership of it. But they had it in virtue of the law of their own city, not of Roman law. Such land was *ager privatus ex jure peregrino*.

These legal differences do not seem to have had any very serious consequences in practice. "The jurists," M. Cuq writes, "afterwards tried to define the relation existing between the holders of provincial land and the State which collected the tithe. They said that the State kept the ownership, while the inhabitants of the provinces only had the possession and a sort of usufruct. But in fact, save for the obligation to pay the *vectigal*, they had all the advantages of ownership. Their right differed from that of full ownership in theory and form rather than in practice and substance."[1] It is true, but the difference clearly shows the origin and meaning of the land-tax among the Romans. That tax did not represent the contribution of each landowner to the expenses of the State, in proportion to the area, value,

[1] XCI, p. 251.

and return of his land. It bore witness that in the past the land of which his estate was composed had been conquered by Rome, and that in the present, in theory if not in fact, the conquering State maintained a right over it. Although arising from another origin and founded on a different basis, this conception of the relations of State and individual led almost to the same result as the organization of agriculture in the Hellenistic East.[1]

The actual forms and distribution of property did not present the same variety as its legal organization. The system of big estates, *latifundia*, still had an important place in the agriculture of Italy, and under the Empire we find it in many provinces in the East and in the West, in Africa, Gaul, Britain, Asia Minor, Syria, and Egypt. The masters of these vast domains were the Emperor himself, either as head of the government and representative of the State or in his private capacity, or members of the Roman nobility, of Senatorial rank, or again, in Asia for example, sanctuaries, like those of the Mother of the Gods at Pessinus and Ma at Comana in Pontos. Some of these great provincial estates had certainly been formed before the Roman conquest (chiefly in the East); there were some, too, which were formed by the concentration in the hands of the Emperor or wealthy Romans of land which had originally been assigned to the public domain.

Whatever may have been the part played by the big estate in the economy of the Empire, medium-sized and small property were the most widespread forms of possession of the soil. The owners of such estates belonged to the population of the towns, to the municipal middle or working class. We cannot draw up statistics, not having detailed evidence; but from the general information supplied by the authors and inscriptions it would appear that the prosperity of the Empire, at least in its first two centuries, was mainly due to the farming of the small and middling rural property. When these two, in consequence of various political and fiscal circumstances, were abandoned or neglected, the agricultural wealth of the Roman world was attacked in its vitals.

To Imperial domains and private property of various

[1] See above, pp. 109 *ff*.

extent and importance, we should add, to obtain a fairly complete picture, common land and also the land left to certain tribes which were still half-nomadic, at least at the beginning of the Empire, in North Africa and several parts of Western Asia.

The methods of farming and the organization of labour to some extent varied according to the various forms of rural ownership.

If we except the Procurators who, in every province or in every administrative district (*tractus*, *regio* in Africa, *diœcesis* in Egypt), supervised the exploitation and conduct of the Imperial domains and had a mainly administrative function, the cultivation of these domains was usually done by tenant farmers (*conductores*), who dealt with the Imperial Procurator and took land of varying size on lease for an agreed rent, and by cultivators (*coloni*) who did the actual work; these last had to pay the *conductor* part of the harvest and do work for him (*partes agrariæ*, *operæ*, *juga*). It is very probable that the *coloni* were, on most of these domains, native peasants, the former inhabitants of the place or their descendants. The relations between them and the big tenant-farmers were not left to the caprice of the latter. Regulations issued by the Emperor determined the amount of the *partes agrariæ* and the number and length of the spells of forced labour. One of these regulations, the *Lex Hadriana*, is mentioned in several inscriptions from Africa.[1] Like the *conductores* and *coloni* in Africa, the μισθωταὶ οὐσιακοί and γεωργοί in Egypt were the chief factors in the exploitation of the Emperor's property. In the great domains of Asia Minor the organization seems to have been rather different; the *colonus* or γεωργός dealt only with the collector of his rent, towards whom he had no particular obligation.[2] In Italy, the Imperial domains were sometimes worked direct by the government, employing the slave labour of a *familia rustica* under the supervision of an *actor*. It was usually so when the domain consisted of extensive pastures, where a few slaves were enough to keep big flocks of sheep.

The big private estate, often formed as the result of an Imperial favour to a high official, a member of the Senatorial order, or a freedman who had managed to secure the con-

[1] **V**, vol. viii, 25943, 26416. [2] **XCIX**, pp. 297 *ff*.

fidence of the master of the Empire, might be worked in a variety of ways. The owner might farm it himself. He might leave the management of it, under his own supervision, to a bailiff (*villicus*), who was almost always a slave. He might lease out all or part to a tenant. Labour was usually supplied, as on the Imperial domains, by local peasants (*coloni*), free-born cultivators whose rent in kind and in labour was determined by regulations (*leges*). The inscription of Henshir-Mettish in Tunisia mentions one of these regulations, the Mancian Law, of unknown date and origin, which seems to have served as a model for several similar regulations regarding big African estates. According to the organization and exploitation of the estate, rent, contributions in kind and labour, had to be paid by the *coloni* direct to the owner (*dominus*), to the tenant (*conductor*), or to the bailiff of one or the other (*villicus*). The formula *domini aut conductores villicive* is repeated several times on the Henshir-Mettish inscription, a kind of charter of the constitution of a big private estate in Proconsular Africa.[1] In Italy, Pliny the Younger had huge estates. According to the circumstances, he leased part of them for a money rent, or he allowed them to be farmed in return for a share of the produce. He speaks in his letters of a *procurator*, *actores*, *exactores*, and *custodes*; that is, of a staff whose business it was to supervise the *coloni*.[2] In Egypt, members of the Imperial family, Senators, Knights, freedmen, wealthy Alexandrians, and afterwards, under the Flavians and Antonines, Greeks more or less tinged with Egyptian blood, held the ground in thousands of *arourai*. The labour for the exploitation of this fertile land was supplied by the Egyptian peasants, whose condition seems to have been no better at that time than in the distant days of the Pharaohs or of the Persian domination.[3]

Elsewhere, the big estate was worked by free tenants, each farming a small area.[4] The great sanctuaries of Asia Minor employed whole armies of slaves on their domains.[5]

[1] **V**, vol. viii, 25902. One frequently hears it suggested that this document referred to an Imperial domain; but to this hypothesis there are many serious objections. To maintain it, fundamental data have to be ignored in the text of the inscription itself. I cannot accept such dangerous methods. *Cf.* J. Toutain, in **XVI**, 1899, pp. 143 *ff.*

[2] Pliny, *Epist.*, iii, 19; ix, 37.

[3] **XCVIII**, pp. 266 *ff.*

[4] *E.g.*, in Syria; *ibid.*, p. 245.

[5] *Ibid.*, p. 287.

The small and medium-sized estates, which were always attached to a town, for a long time owed their prosperity to the intelligence and hard work of the municipal middle class. In many provinces of the Empire, the country round the towns was covered with villas and farms, houses for pleasure and buildings for work. The owner of the estate doubtless came and stayed there in the good season, either to enjoy the country life or to supervise the harvest, vintage, olive-picking, and manufacture of wine and oil. A bailiff, a few slaves, and sometimes also free labourers, taken on either by the year or for some particular urgent work, such as the harvest, were enough for the exploitation of these modest estates. Sometimes, too, they were leased for a money rent, or to *métayers* who paid a proportion of the produce. A few passages in the ancient authors, inscriptions, and mosaics have revealed the character, at once static and dynamic, of the small and medium estate in Italy and in some of the provinces. Horace's villa in the Sabine country "was not merely the little garden of a literary man, a lizard's hole, as Juvenal says. . . . It was a real estate, with meadows, fields, woods, and a whole farm, a source of income as well as of pleasure."[1] Of the villas excavated round Pompeii, that of Boscoreale was certainly surrounded by a property which was farmed, for its ruins have yielded up presses, underground channels intended to collect the wine, *dolia* sunk in the ground, apparatus for the manufacture of oil, and the floor where the grain was threshed, and the dwelling-house of the owner and the lodgings of the workers have been identified and distinguished.[2] Here, therefore, there were cornfields, vineyards, and olive-groves, and oil and wine were made on the spot.

For Africa we have several documents which give us a notion of what an estate of average size was like. Near Tabarka, on the shore of the Mediterranean, there is a celebrated mosaic in three scenes, representing a rural property. One picture shows the dwelling-house and park. "The house," Monsieur P. Gauckler writes, "seems to be built on the plan of the modern *borj*. At the back of a square court is the master's house, a one-storied building flanked by two

[1] G. Boissier, *Nouvelles Promenades archéologiques*, p. 36.

[2] H. Thédenat, *Pompéi : histoire, vie privée*, p. 155; A. Mau, *Pompeji in Leben und Kunst*, pp. 356 *ff*.

tall square turrets with pointed roofs, which are connected half-way up by a loggia. In front are the coach-houses and offices. . . . On the left is a barn, surmounted by a dove-cote with three windows. In front of the *borj* is a pond, with ducks and geese playing about on the edge. At the sides and behind is a flowery orchard full of pheasants and other birds." Another of the three pictures deals more specially with the vineyard and orchard and the buildings needed for working them. The third shows us the pastoral part of the estate—sheep, a shepherdess sitting spinning with her distaff, the stables with a horse tied to the door, and the uncultivated hills behind.[1]

South of Carthage, in the lower course of the Wadi Miliana on the site of the ancient town of Uthina, a mosaic has been found which shows, in several episodic scenes, the chief features of daily life on a medium-sized property. In addition to various hunting-scenes—hunting in the ordinary sense, hunting with the net, hunting with the spear—we see a ploughman guiding a plough drawn by two oxen, a flock of sheep returning to the stall, by the door of which a shepherd is standing, a horse drinking at a trough near a swing-well, a slave leading an ass, perhaps to market in the town, another slave picking olives, a third milking a goat, a fourth playing the flute while watching his beasts. Here corn-growing, stock-breeding, and tree-growing are represented as it were in synthesis.[2]

In the south of Proconsular Africa, the estate formed by T. Flavius Secundus, doubtless one of the colonists settled by Trajan on the territory of Thelepte, consisted chiefly of vineyards. The irrigation which the owner carried out must have been good for garden produce and trees.[3]

The life and career of a small landowner in the same province are described clearly in the epigraphic text known as the *Inscription of the Harvester of Maktar*.

"I was born of a poor family. My father had no income or house of his own. Since the day of my birth, I have always tilled my land; my land and I have not rested. When the season of the year came

[1] *Inventaire des mosaïques de la Gaule et de l'Afrique*, vol. ii, p. 303; atlas, no. 940.

[2] *Ibid.*, p. 122; atlas, no. 362.

[3] J. Toutain, *Les Cités romaines de la Tunisie*, p. 318; V, vol. viii, no. 212, p. 1, ll. 51 *ff.*

round when the harvest was ripe, I was the first to cut my stalks. When the bodies of harvesters who hire themselves out round Cirta, the capital of the Numidians, or in the plains overlooked by the Mountain of Jupiter, appeared in the country, then I was the first to reap my field. Then, leaving my country, I harvested for other men for twelve years under a burning sun. For eleven years, I commanded a gang of harvesters and reaped corn in the fields of the Numidians. By my work, having been able to do with little, I at last became the owner of a house and an estate. Today I am comfortably off. I have even risen to honours; I have been inscribed among the Decurions of my city, and my colleagues have elected me Censor, me who at the beginning of my life was only a small peasant. I have seen my children and grandchildren come into the world and grow up round me, and my life has gone by, peaceful and honoured by all."[1]

At first a tenant farmer or *métayer* of modest condition, then a farm-labourer, then a contractor for harvesting, and lastly a landowner—these seem to have been the stages of that hard-working existence. And the case of the Harvester of Maktar was probably not unique or even exceptional in the African provinces.

In Gaul, where the remains of villas and farms are very numerous, medium-sized and, above all, small property seem to have had a considerable place, " especially on the outskirts of towns."[2] Of the rural life and activity of the Gallo-Romans we have evidence in the rustic calendar represented in the mosaic of St. Romain en Gal on the outskirts of Vienne[3] and in various monuments adorned with agricultural scenes or motives—ploughing, vintage, gardening-tools, etc.[4]

So it was, doubtless, in all parts of the Roman Empire where wealth was chiefly agricultural, in Bætica and eastern Tarraconensis, in Dalmatia, in Asia Minor, in Egypt. The medium and the small property naturally do not take such a place in written documents, legal or epigraphic, as the big private or Imperial estates. What we know of their importance and the part which they played in the economic life of provinces like Africa, authorizes us to suppose that these were similar wherever municipal life developed, wherever there were many thriving towns.

Lastly, we must mention a special form of occupation and working of the soil—the emphyteusis. " The emphy-

[1] V, vol. viii, 11824. [2] XCVI, vol. v, pp. 362 *ff.*

[3] *Inventaire des mosaïques de la Gaule et de l'Afrique*, vol. i, p. 54; atlas, no. 246.

[4] Espérandieu, *Recueil*, nos. 464, 1682, 2793, 3114, 3472, 3478, 3681, 4092, 4243, 5619, 5833, etc.

teusis was an agreement, the principal object of which originally was to bring uncultivated land on the Emperor's private domain under culture. It was afterwards extended to inherited estates, those of churches and private persons. The name of this agreement comes from the obligation imposed on the taker of planting (ἐμφυτεύειν) or at least improving the ground. . . . It was a long-term lease, which had its origin in certain practices which had long been followed in the administration of large estates. In the first centuries of the Empire, the landlords allowed their *coloni* to occupy uncultivated portions in order to reclaim them. In return, they were given the exclusive enjoyment of the fruits for a certain number of years, after which they had to pay a small rent in kind. Moreover, they were allowed a sort of real right over the thing, which was revocable when they ceased to cultivate the land for two years."[1] The most typical case of a deed of emphyteusis in the Early Empire seems to be the decision of Pertinax, which is thus described by Herodian: "Pertinax allowed anyone who could and would to occupy all land in Italy and the provinces which was not under cultivation at the time or had never been cultivated, even if it belonged to the Emperor, and he ceded the ownership of it to those who occupied and worked it in this fashion. He granted these cultivators exemption from all taxes for ten years, and guaranteed them the perpetual ownership of the land in question."[2] There might be various forms of emphyteusis. The essential basis of the right thus created in favour of the cultivator was the fact that he brought under culture land which had previously been neglected and left unproductive.

In whatever variety of manners property may have been organized and the soil worked, what emerges from the facts at present known is that, so far as the civilization of the time and the historical accidents of the Roman Empire allowed, most of the countries united under Rome enjoyed an agricultural and pastoral fertility which many of them had not known before, which some, such as North Africa and Egypt, have recovered only after long centuries, and for which others, such as Asia Minor, are still waiting.

The share of credit for this happy result which one may

[1] XCI, p. 357. [2] ii, 4, 6.

fairly assign to the sum of conditions covered by the words *Pax Romana* is the greater in that the material equipment of agriculture does not seem to have been much improved at the time. For grinding corn, the mill driven by beasts (*mola asinaria, jumentaria*) or even by water (*mola aquaria*) more and more took the place of the old pestle and mortar. The press and other apparatus for the manufacture of wine were improved. Olive-mills and oil-presses were more workable and gave better results.[1] But the real implements of the farmer still had their old shape. Plough, spade, hoe, mattock, pick, fork, scythe, sickle, and pruning-knife were, as the surviving specimens show, just as they had been handed down from generation to generation.[2] Nor was there any more change in the methods of stock-breeding; an African mosaic representing shepherds guarding their flocks might serve as an illustration to Theocritos, if not to the *Odyssey*.

[1] **XX**, vol. ii, pp. 230 *ff*.
[2] S. Reinach, *Catalogue illustré du Musée des Antiquités nationales au château de Saint-Germain-en-Laye*, vol. i, pp. 273 *ff*.

CHAPTER III

INDUSTRY AND INDUSTRIAL LABOUR

THE general conditions of public and private life influenced industry and the organization of industrial labour as much as they did agriculture, stock-breeding, and the organization of property and farming.

In the Eastern provinces, except in Greece, whose decay was past curing, the establishment of the Empire and the advent of the Roman Peace and political and administrative unity gave a new impulse to the many industries which we have already seen thriving in the Hellenistic period.[1] The extension and development of trade between the Mediterranean world and the Far East, India, and East Africa; the increasing demand in the West for all the products of the luxury industries of Alexandria, Phœnicia, Syria, and Asia Minor; the safety of communication by sea—all these consequences of Roman rule helped greatly to increase the output of the workshops of the East. But those workshops had already been long in existence; in the case of the East, we cannot speak of a profound transformation or even of a very marked development.

In the West, on the other hand, industry, in a thousand different forms, went through two or three hundred years of intense and fruitful life. The causes of this economic phenomenon were many. The foundation of numerous towns in North Africa, the Iberian Peninsula, Gaul, and the Danubian countries; the creation of an admirably co-ordinated system of land-roads and the organization of inland navigation, which facilitated the relations of one province with another and therefore the circulation of manufactured goods; the increasing demands of the consumers, not only local consumers, but those of the surrounding region and even of other regions; the discovery and increasing use of raw materials which had previously been little known and inefficiently

[1] Above, pp. 122 *ff*.

utilized—these were the principal facts, political and economic, to which the western and northern provinces of the Roman Empire owed either the birth or the progress of their industrial production.

Whatever differences one may perceive between East and West, between the Greek and Latin provinces of the Empire, the industry of antiquity, which had once been dispersed, was now to some extent concentrated. Some kind of balance was established between the countries with a highly developed civilization, in which industry had been going on for hundreds of years, and those which had, down to the Roman conquest, remained more or less cut off from the Mediterranean world. From one end of the Empire to the other, the products of industry went about unceasingly, and production increased as consumption became more general and outlets and markets became more numerous.

I

Raw Materials and the Extractive Industries: Salt-Works, Quarries, and Mines

As they themselves progressed, agriculture and stock-breeding furnished industry with more abundant and varied raw materials—corn, grapes, and olives for the manufacture of foodstuffs, flax, hemp, and wool for weaving, and hides for tanning and leather-work. The building-trades could obtain wood for heavy carpentry and joinery in forests which were more and perhaps better exploited. Sea-fisheries and shell-fisheries continued to supply the salting and dyeing industries respectively.[1]

It was chiefly the raw materials of mineral origin which took the first economic place in the Roman world. First of all, the development of town life, which marks the evolution of many parts of the West and Central Europe in this period, was necessarily attended by the seeking out and employment of building-materials—stone of all kinds, lime, plaster, sand. M. Jullian has said, with reference to Gaul: " For three centuries, the ordinary products of the soil were continually

[1] *Cf.* above, pp. 261 *ff.*

collected in cartloads—the limestone, sand, and plaster which were used for the wall-mortar and floor-concrete, of which the Romans consumed such quantities for their buildings and roads, and gravel and small stones. . . . Lastly, more important than everything else to the history of the soil of Gaul, was the opening up of quarries of building-stone. . . . Never in all its life has the ground of France been subjected to more digging, study, and analysis than in the centuries of the Roman Empire. Nothing was left to chance. When a large building was being erected, different quarries were used, each for a special purpose. One supplied the facing-stones, another the masonry of the vaults, another the rubble filling. For three hundred years the quarryman's pick clinked unceasingly from one end of Gaul to the other."[1] Potter's clay was used quite as much. "Every city," says M. Jullian again, "had its own tile-works and potteries, if it was only for building-bricks and the commonest utensils, and it made arrangements to obtain its raw materials on the spot."[2] What is true of Gaul is equally true of Africa, where towns rose from the Numidian soil in great numbers; of the Peninsula, where Bætica, Tarraconensis, and even some parts of Lusitania were the scene of intense town life; of Britain, where the first cities worthy of the name were now founded; and of the military frontiers on the Rhine and Danube, where the great permanent camps established by the Romans already marked the sites of the cities which are now called—to mention only the chief of them—Cologne, Mainz, Strasburg, Ratisbon, Vienna, Budapest, Belgrade, and Silistria. If we think of the enormous amount of stone, building-materials, bricks, tiles, and crockery needed for the creation, growth, and daily life of all the towns which came to cover the Latin provinces of the Empire, we shall have an idea of the activity then expended, from the edge of the Sahara to the borders of Germany and from the Atlantic to the Euxine, wherever the earth could supply the necessary raw materials.

Of the extractive industries, some, which were of a more special kind, deserve particular attention, by reason both of their nature and of their organization.

The ancients knew how to obtain salt both from salt marshes by the sea and inland from salt lakes, hot springs,

[1] **XCVI**, vol. v, pp. 213 *ff.* [2] *Ibid.*, p. 211.

and mines of rock-salt. The best-known centres of this industry were: in Italy, near the mouths of the Tiber round Ostia, on the territory of Volaterræ in Etruria, and in the environs of Tarentum; in Sicily, near Gela and Agrigentum (Acragas); in Africa, round Utica and in some districts on the Egyptian side of Libya; in Nearer Spain, at Egelasta, north of New Carthage (Cartagena) in the valley of the Sucro; in Gaul, on the coast of Languedoc, among the Tarbelli of Dax, along the Atlantic to Flanders, and among the Sequani and Mediomatrici; in the Balkans, in Epeiros, Illyria, and Thessaly; in Asia Minor, round the Anatolian lakes of Phrygia and Cappadocia; in the island of Cyprus; and in many parts of Egypt.

Various texts tell us how the salt was extracted, according to its origin and the local conditions. This is the account which Rutilius Numatianus gives of a salt marsh near Volaterræ:

> "The sea enters upon a slope by channels dug in the ground, and a small stream fills innumerable pools. But when the Dog-star comes with his burning fires, when the grass fades and all the land is athirst, then the sea is shut out by the barrier of sluices, so that the water, held fast, is hardened by the hot ground. Under the lively influence of Phœbus, the elements coagulate into a thick crust."[1]

This is exactly the method of evaporation used in the salt marshes at this day.

Rock-salt was hewn from the mine in blocks. Strabo says of Lake Tatta that it may be compared to a natural salt-pit:

> "The salt contained in its water sticks so readily to anything which one dips in it, that if you put in a hoop of reeds you draw out a crown of salt."[2]

The processes by which salt was extracted from saline springs varied in different countries.

Among the quarries, those of marble, porphyry, and some kinds of granite were particularly important. The Roman love of luxurious materials and splendid buildings explains why blocks of marble, granite, and porphyry of different colours, dark or brilliant, were unloaded on the wharves of the

[1] i, 475 *ff.*
[2] xii, 5, 4. For salt-works in the Roman Empire, see M. Besnier, in **XVII**, *s.v.* "Sal."

Tiber at the foot of the Monte Testaccio. Here again, the appearance of towns and the development of town life in many provinces led to a more active working of the old quarries of Greece, Asia Minor, and Egypt and the opening up of deposits previously unknown or neglected in North Africa, Gaul, and the countries on the Rhine and Danube. In the first centuries of the Empire, Italy supplied the white marble of Luna and the grey granite of the island of Elba. In Greece, work was continued or revived in the famous old marble-quarries of Pentelicon and Hymettos in Attica, Carystos in Eubœa, and Paros, Thasos, and other islands of the Ægean, and green porphyry was extracted on the peninsula of Tænaron. The workings of Synnada in Phrygia were very active. Egypt produced the red granite of Syene, the red porphyry of Mons Claudianus (Gebel Dukhan), and the alabaster of Mons Berenicidis. North Africa supplied, among other marbles, the yellow antique of the quarries of Simitthu (Shemtu), the red of Ain Smara near Sigus, the white of the Gebel Filfila west of Bona, and the onyx-alabaster of Ain Tekbalet in the district of Oran. In Gaul, the Pyrenæan marbles of St. Béat, Marignac, and Barousse and the granite and marble of the Vosges were already renowned. The quarries of purple granite on the Felsberg in Hessen-Darmstadt and those of porphyry on the Frushka-Gora north-west of Belgrade, between the Drave and the Save, were now opened for the first time. In every part of the Empire men laboured at the extraction of materials possessing the colours and qualities which were in demand.

Traces of ancient workings have been observed and studied on the sites almost everywhere. The processes revealed—open quarrying, the boring of underground galleries, the hewing of long rows of blocks in successive terraces, the rough-hewing of blocks intended for columns in the quarry itself, before they were detached, and the use of wooden wedges to split the rock—have enabled us to reconstruct the appearance of an ancient quarry with certainty.[1]

The deposits of metallic ores, both of the precious metals, gold and silver, and of the base metals, iron, copper, lead, and tin, did not have the same history in Greece and the

[1] C. Dubois, *Étude sur l'administration et l'exploitation des carrières dans le monde romain*, Paris, 1908; **XX**, vol. i, pp. 8 *ff*.

East as in the western and northern parts of the Roman world. The mines of Greece were almost exhausted, and Asia Minor produced hardly any more gold. Macedonia and Thrace were still rich in gold and silver; Colchis and Egypt continued to produce gold; and the iron of the country of the Chalybes, the copper of Macedonia and, still more, of Cyprus, and the lead of the Troad and Cilicia had not ceased to be worked. But these regions were far outdone by the West and North in the abundance, variety, and richness of their ores. Gold was obtained in various forms, nuggets and float gold, in Spain, Gaul, Britain, Dalmatia, and Mœsia; Trajan, by his conquest of Dacia, presented Rome with one of the richest goldfields in the ancient world. The mines of argentiferous lead in Sardinia, all the region of the Sierra Morena in Spain, central and southern Gaul, Dalmatia, Pannonia, and Epeiros furnished the Roman world with considerable resources. The metal industry was supplied with iron by Elba, Etruria, the parts of Spain near the Pyrenees, many deposits all over Gaul, and especially the mines of Noricum (now Styria and Carinthia). Bætica and Lusitania in the Peninsula, Aquitania and Narbonensis in Gaul, and the island of Britain competed with Cyprus in supplying the Empire with copper, and the same regions, by their production of tin, enabled the bronze industry to make great strides everywhere. North Africa and Britain produced lead.

No doubt, the mines of southern Spain had already been known and worked for a long time, and the almost fabulous renown of the land of Tartessos shows what an impression their wealth had made among the peoples of the Eastern Mediterranean. No doubt, too, the Gauls had practised many forms of metal-working before the Roman conquest. But it was in the early centuries of the Empire that the metallic resources of Western and Central Europe, from the Pillars of Hercules to the lower Rhine and from Britain to Transylvania and the Balkans, were really made productive. The fact is not only proved by the evidence left us by the past in the shape of literary and historical works, inscriptions, still visible traces of workings, and stamped pigs and ingots of various metals; it is manifest from the countless heaps of ancient slag, from which modern miners are able, with their improved processes, to obtain considerable quantities of metal.

Moreover, the Imperial government very soon saw the importance of these riches and the part which they should play in the economic life of the Roman world. Its attention was attracted not only by the mines of precious and other metals, but by the quarries of luxury stone, such as marble, porphyry, and certain granites, and the marine and inland salt-beds. It would be excessive and therefore untrue to say that mines, quarries, and salt-works were legally the property of the Emperor. In the second century of the Empire, under the Antonines, the marble-quarries on Pentelicon belonged to the celebrated rhetorician Herodes Atticos. This, however, seems to be an exceptional case. If not legally, at least in practice, mines, quarries, and salt-works, or at least those of any importance, were part of the Imperial domains.[1] Some came from the public domain of Republican times; others had come into the hands of the Emperors by purchase, deed of gift, inheritance, confiscation, or conquest. To all those existing in the countries which were made provinces before the accession of Augustus, in Gaul, Dalmatia, Africa Proper and Numidia, Greece, Asia Minor, etc., Augustus and his successors had added the rich deposits of Britain, Noricum, Pannonia, Dacia, and Egypt. This was a source of considerable wealth for the Imperial government.

There is, however, no trace in Rome itself of a central administration of these industries, at the head of which there would have been a *Procurator Metallorum*. It is usually believed that they came under the *Procurator Patrimonii*. Under the direction and control of that high official, junior Procurators had the management either of a single important mine, quarry, or salt-works, or of a group of mines, quarries, or salt-works in a particular district. Thus we hear of special Procurators of the marble-quarries of Simitthu in Africa and Synnada in Asia Minor, of the granite-quarries near Syene and the quarries of red porphyry on the Mons Claudianus in Egypt, and of the copper-mines of Vipsaca in Lusitania; and we also hear of provincial or district Procurators of the gold-mines of Dacia, the silver-mines of Pannonia and Dalmatia, and the iron-mines of Noricum.[2]

[1] **XXI**, vol. x, pp. 326 *ff.*; *cf.* pp. 317 *ff.*, and Ardaillon, in **XVII**, *s.v.* "Metalla," p. 1871.

[2] Ardaillon, *loc. cit.*; C. Dubois, *Étude sur l'administration et l'exploitation des carrières, passim.*

The function of the Procurators was purely administrative. The actual working of the mines and quarries was done either by lessees or by the government itself. At Vipsaca, as we know from the *Lex Metalli Vipsacensis*, the former system was employed. The Procurator farmed out the exploitation of the vein to one or more tenants (*conductores*), who themselves entrusted the work to contractors, who were no doubt specialists in various operations—extraction of the ore, smelting, etc. There are traces of the same organization in the iron-mines of Noricum, the quarries of Egypt, and many salt-works, such as those round Ostia. Elsewhere the exploitation was done under the direct supervision of the Procurator, by a technical staff mainly composed of freedmen and Imperial slaves. This appears to have been the case in the gold-mines of Dalmatia, Pannonia, and Dacia and the marble-quarries of Carystos, Synnada, and Simitthu.

Are we to believe, as has been suggested, that the system of leasing out was employed for mines, quarries, and salt-works belonging to the *fiscus*, that is, the public domain properly so called, and that of direct exploitation for those which were regarded as the personal property of the Emperor?[1] The question is not of great importance, for the distinction of the *fiscus* from the *res privata* and *patrimonium Cæsaris* was much more theoretical than real.

So the Imperial government had secured, in addition to its huge rural estates, mineral resources of exceptional abundance and richness. It cannot be said that it had established a monopoly of the exploitation of these resources for itself, for the Digest speaks of quarries, mines, and chalk-pits which are private property.[2]

Great as were the abundance and variety of the raw materials which industry obtained from the provinces of the Empire, the Roman world imported some from abroad—amber from the shores of northern Germany, alabaster from the kingdom of the Sabæans in southern Arabia, ivory from Africa, tortoise-shell from Africa and India, certain luxury woods, such as ebony, teak, and sandal, from India and Ethiopia, hides from Africa and the Iranian plateau, a textile product

[1] Ardaillon, *art. cit.*, p. 1872.
[2] Bk. xxvii, tit. 9, 3, 6.

which was called *byssus* or *lana lignea* by the ancients and is sometimes supposed to have been a kind of cotton from India, raw silk from China, and iron from Iran.[1]

II

Industrial Production. The Principal Industries

Apart from the general conditions indicated above, which contributed to the development of industrial production in the Roman Empire, the amount and variety of the raw materials exploited in the first centuries of our era gave a lift to industry which it had never had before, at least in the West.

First of all, we should note, without dwelling too much on it, the creation of many purely local industries in all the towns which were founded in Africa, the Peninsula, Gaul, and the Rhenish and Danubian provinces. Every one of these towns had economic needs, which were met by the output of little industrial concerns which had no ambition to do business on a large scale. In every town local consumption and that of the surrounding district required the organization of a minimum of production and labour.

In the East, if the Roman Peace encouraged the progress of industry, it cannot be said that in this respect there was any revolution or even a transformation. The industrial activity of Egypt, Syria, and Asia Minor changed neither in character nor in direction. Alexandria, Tyre and Sidon, Antioch, Miletos, and the Euxine continued to manufacture and export transparent or purple-dyed tissues, artistic bronzes of every kind, glassware, scents, smoked or salted fish, and so on.[2]

The western and northern provinces, on the other hand, underwent an absolute metamorphosis. This does not mean that industry now appeared there for the first time. We have seen that before the Roman conquest those regions had given proof of initiative, skill, and experience in this domain.[3]

1 **XCVII**, pp. 293 *ff.*; R. Cagnat and M. Besnard, in **XVII**, *s.v.* " Mercatura," pp. 1773, 1778.

2 Above, pp. 122 *ff.*

3 Above, pp. 175 *ff.*, 180 *ff.*

But the actual organization of the Roman world, by assisting and multiplying relations between all countries near the Mediterranean, by widening the horizon to which the activity of each of those countries could extend, by encouraging the exportation of its manufactures, sometimes to distant markets, enabled certain industries of the West and North to expand, and therefore to produce on a scale to which they had not been accustomed.

Among these industries, we should observe especially those of building and furniture-making, metal-working, pottery, weaving, and luxuries.

The prosperity of the building-industries was the inevitable result not only of the spread of town life and the prosperity of the *municipia*, but of a general improvement. Stone-buildings took the place of the *mapalia* of Africa and the wattle-and-daub huts of Gaul. Regions which were formerly uninhabited or left to nomads were covered with luxurious villas and rich, well-equipped farms. Between the towns, which sometimes stood far apart, thriving villages of varying size appeared almost everywhere. Masons, stone-cutters, and heavy carpenters were the chief creators of that achievement, in many places a truly monumental achievement, and covered the West "with a white garment of edifices," the vestiges of which still inspire astonishment and admiration. The famous Roman cement, composed of lime, sand, and brick-dust caused these buildings to endure for centuries.[1]

With these new conditions of housing went a corresponding improvement in the furniture used in daily life. Like the building-trades, the manufacture of furniture became remarkably active and prosperous. Beds, seats, tables, cupboards, chests, candlesticks, and lamp-stands, most of them doubtless wooden, but some made of bronze, marble, or stone (which are the ones which have survived), filled the rooms of the houses.[2] To realize the progress which must have been made in all these branches of furniture-making, it is enough to compare what the Numidian and Gallic huts and the cabins of Pannonia or Mœsia must have been like with the houses of the Roman period whose ruins have been found in North Africa, France, Belgium, England, the Rhineland, and the

[1] **XCVI**, vol. v, pp. 223 *ff*.
[2] **XX**, vol. ii, pp. 409 *ff*.; *cf*. pp. 464 *ff*.

Danubian countries.[1] What we call comfort—but it was still a very comparative comfort—was brought into the everyday life of the peoples of the West, and to provide that comfort many industries were necessary.

The various forms of metal-working had long been known and practised in the West, principally in the southern part of the Peninsula, Gaul, and certain districts in Italy, such as Etruria and Campania. Here there was progress not so much in the technical processes as in the extension of the industry, in the number of workshops built and the number and extension of the markets opened to it. In Spain, not only were iron and copper ores extracted from the rich veins scattered about the valleys of the Guadalquivir and Guadiana; those ores were transformed on the spot into workable metal, iron, steel, bronze, and the metal was manufactured into swords, daggers, breastplates, and other military articles. Bilbilis, the home town of the poet Martial, was renowned for its arms, and the bronze of Cordova was in especial demand.[2] Gallic metal-working, which had been so active when the country was independent, far from suffering by the Roman conquest, seems to have benefited by the new régime. "The metal industries," M. Jullian writes, "had been the liveliest when Gaul was free. They were very greatly altered by the conquest, which introduced all, manufacturers and customers, to the great international markets of raw materials and manufactured goods."[3] The bronze-workers made quantities of fibulas and ornaments of all kinds, statuettes, vases, and handles and fittings for furniture and chests. "In Gaul, bronze-working set the crown on its triumphs of the last two thousand years."[4] It was above all iron-working that developed in Gaul. "A whole world of workers was employed on it, from great, wealthy ironmasters to humble country blacksmiths. Hammers and tongs perhaps appear on the tombs more frequently than any other tools. No industry did more for the life of the Gauls, their real life, that of the camp, the land, the workshop, and the home. It provided the peasants with ploughs, harrows, scythes, knives, axes, and pruning-knives; the carpenters and masons with hammers, chisels, pincers, saws, nails of all shapes, nuts,

[1] **XX**, vol. i, pp. 295 *ff.*
[2] **XXXV**, pp. 127 *ff.*, *passim.*
[3] **XCVI**, vol. v, p. 300.
[4] *Ibid.*, p. 306.

wrenches, files, rules, compasses, anvils, and the famous iron axes or *asciæ* which are so frequent in the funerary symbolism of the Celts; the warriors with their arms of all kinds, particularly their swords, the trade in which alone sufficed to keep a business house going; and the hunters with knives, darts, and spears. . . . In household life, locksmith's work was then at its best, with its fastenings, bolts, staples, padlocks, and keys with various wards, some of which last were enormous, as complicated as machines and as heavy as weapons."[1] In no country do the museums offer a more complete and varied collection of ancient iron tools and implements for observation and study than in France.[2]

Noricum, which possessed the richest iron-mines in the Roman world, became a very busy industrial centre in the first centuries of the Christian era. The products of its forges, especially the swords, were renowned even in Italy. At the end of the Empire, the iron of this region was supplying several arms-factories on the Danubian frontier: Lauriacum (Lorch), Carnuntum (Petronell, east of Vienna), Aquincum (Budapest), and Sirmium (Mitrovitza, on the Save, west of Belgrade).

In Italy, the chief centres of the metal industry were still in Etruria and Campania. The island of Elba, Populonia in Etruria, and Capua in Campania manufactured iron and, still more, bronze goods. No doubt, the foundries and small-iron works of Etruria had lost some of their old brilliance and prestige; but contemporaries of Augustus, for example Strabo, still speak of the activity of the workshops of Populonia, which received the metal direct from Elba. In the time of Pliny the Elder, Campanian bronze was believed to be the best for the manufacture of vases, basins, cooking-pots, etc.[3]

Like metal-working, pottery, in its many forms, conquered a chief place in industry in the time of the Empire. But it was not an artistic industry comparable to that which had once presented the ancient world with the painted vases of Corinth and Athens and the statuettes of Tanagra, Myrina, and Tarsos. Tile-works and brickfields naturally sprang up in great numbers in the western provinces, as stone houses

[1] **XCVI**, vol. v, pp. 307 *ff*.
[2] For Belgium, *cf*. **XC**, pp. 37 *ff*., 74 *ff*.
[3] **XXXV**, pp. 108, 116.

roofed with flat or curved tiles (*tegulæ*, *imbrices*) took the place of the old huts of wood, reeds, and the like. Apart from them, we chiefly know the pottery of Imperial times by household ware, often decorated, no doubt, but intended for commonplace purposes, and statuettes whose clumsiness, awkwardness, and commonness are patent to the kindest critic. It is obvious that the public for which these vessels and statuettes were intended did not display the enlightened, delicate, exacting taste of the Greeks of the days of Solon, Pericles, or Demosthenes. The pottery of the first centuries of our era reveals an industry which was chiefly anxious for a large output, and manufactured wholesale.

The chief centres of this industry were in Italy and Gaul.[1] The workshops of Mutina, especially that of the lamp-manufacturer Fortis, were celebrated all over the Roman world. The Campanian potteries of Cales and Cumæ turned out countless jars, *dolia*, plates, bowls, and dishes. Until about the end of the first century, Arretium was acknowledged supreme in the manufacture of a red glazed ware, sometimes adorned with scenes or motives in relief; later, the competition of Gallo-Roman pottery deprived it of its markets beyond the Alps. The production of Gaul developed wonderfully in the second and third centuries; all over the country, from the Pyrenees to the Rhine, clay was turned, modelled, and decorated with moulded or applied motives or, more rarely, painting.[2] Like the master-potters of Arretium, those of Gaul sometimes signed their goods, and through these trade-marks stamped on the bottoms of vases we now know which potbanks enjoyed the biggest custom. Some were collected in the centre, among the Gabales of the Gévaudan (Banassac), the Ruteni (La Graufesenque, Montans), and the Arverni (Lezoux), and others worked in the east and north, among the Remi and Leuci of the Argonne, the Nemetes and Triboci of Alsace (Rheinzabern, Ittenweiller, Heiligenberg), and the Treveri, Tungri, and Nervii (Treves, Bavay). The manufacture of terra-cotta statuettes, so divine as to be almost shapeless, was as prosperous as that of vases; the potteries

[1] *Ibid.*, pp. 100 *ff.*, 115 *ff.*, 140.

[2] **XCVI**, vol. v, pp. 264 *ff.* *Cf.* J. Déchelette, *Les Vases céramiques de la Gaule romaine ;* Forrer, *Die römischen Terra-sigillata Topfereien in Elsass*; G. Chenet, *L'Atelier du Pont des Rèmes ;* **XC**, pp. 65 *ff.*

from which "these poor works, made for poor households" came seem to have been those conducted by Allusa at Bordeaux, by Rextugenos in Brittany, by Sacrillos among the Arverni at Toulon-sur-Allier, and, last but not least, by Pistillus, probably near Autun, among the Ædui.[1] There may have been similar works among the Tungri of Belgium.[2]

In Spain, Saguntum manufactured cups (*calices*) which were in demand even in Rome;[3] specimens found near the Spanish city lead one to suppose that the technique and decoration of these vases recalled those of the Arretine ware. In the African provinces, and even in Britain, this industry, although it did not develop as it did in Gaul and Italy, was sufficiently prosperous and supplied local needs.[4]

With the ceramic industries, the textile industries—weaving, dressing, dyeing, and clothes-making—seem to have been the most widespread and prosperous in Italy and the whole of the West. Certainly the use of wool and flax and the many transformations of those materials into clothing, blankets, carpets, and stuffs of every kind and for every purpose were not unknown before. Before the Roman conquest, the inhabitants of Cisalpine Gaul and the present France, the peoples of the Peninsula, the Britons, the Africans, and those who dwelt on the Danube and its southern tributaries, however barbarous we may suppose them to have been, were able to clothe themselves with other things than the skins of beasts. But what marks the Imperial epoch seems to have been the localization of the textile industries, or at least a great increase of importance in certain of these industries in a few definite regions, and also a process of specialization, at least to some extent, by which the manufacture of certain materials became as it were the monopoly of fairly restricted populations or districts.

Sheep-farming being practised everywhere, the manufacture of wool and cloth was very active everywhere. In this again Italy and Gaul stood in the front rank. The dark wool of Pollentia, near the Tanaro in Liguria, the woollen goods of Parma and Mutina (Modena) in the present Emilia, and the cloths, carpets, and clothing of Patavium (Padua) in Venetia were renowned among all products of the kind.[5]

[1] **XCVI**, vol. v, p. 287. [2] **XC**, p. 86 n. 4. [3] **XXXV**, p. 132.
[4] **XCVIII**, pp. 163, 538 (37). [5] **XXXV**, pp. 99 *ff*.

In the south of Italy, Luceria, Canusium, and Tarentum worked the wool supplied by the flocks of Apulia, Calabria, and Bruttium. The many mills in and about these cities turned out fine and coarse materials, which were manufactured into the *pœnulæ*, *birri*, *tunicæ russæ Canusinæ*, *Tarantina*, and *Tarantinidia* so often mentioned by the authors.[1] Under the Empire, "Gaul became the biggest cloth-producing country in the world. 'It is she that dresses us common folk,' a poet of the capital said."[2] The Atrebates and Nervii in the north, the Santones in the west, and the Lingones in the north-east made quantities of dark-coloured cloaks and strong ones with hoods. Rouen, Amiens, and Rheims in Belgica, Bourges in the heart of Aquitania, and Nîmes in Narbonensis[3] wove celebrated cloths. Woollen carpets and blankets were also made in Gaul. Flax and hemp were worked no less than wool. The Cadurci were known even in Italy for their linen, and the tribes dwelling along the rivers supplied the ship-owners and fishermen with sails and rigging.

Outside Italy and Gaul, Africa used up the wool of its many flocks itself. Malta specialized in making light materials called *othonia*, but whether from a very fine flax or from a species of cotton is uncertain. In Spain, Tarraco (Tarragona) exported its *carbasi* or *carbasa*, which were cottons of some kind, and the towns of Emporiæ and Sætabis in the same country had a great reputation for the fineness and softness of their linens.

This development of the textile industries in many parts of the West led to a spread of the fuller's trade, which included not only "the fulling of woollen material which was to be made into cloth, but all the operations and handling connected both with cleaning and dressing new cloth and with restoring the condition of what had already been worn as clothing."[4]

These were the principal industries which rose to great prosperity in the West under the Roman Empire, while the East continued on the ways opened to it by the conquests of Alexander the Great. The luxury industries were still active at Alexandria, in Phœnicia, at Antioch, and in some cities in Asia Minor. The industrial development of the West had

[1] *Ibid.*, pp. 121 *ff.*
[2] **XCVI**, vol. v, p. 240.
[3] Rotomagus, Samarobriva, Durocortorum, Avaricum, Nemausus.
[4] A. Jacob, in **XVII**, *s.v.* "Fullonica," p. 1849.

in no way injured that activity. The only result of it was that there was now a better balance between the economic importance of the East and that of the West. The unity of the Empire and the Roman Peace had contributed mightily to that result.

III

The Organization of Industrial Labour

Although industry had so developed in the first centuries of our era, it had not altogether deserted the household. Not only in more or less isolated country dwellings, but even in the towns, bread was still made in the house, where the grain was crushed in hand-mills. The great number of primitive querns, consisting of a flat stone, slightly convex (*catillus*) or concave (*meta*), found in almost all ancient ruins, shows that old customs survived even when baking had become a public industry.

So, too, wool and flax were still spun and woven at home. Fulling and dyeing were perhaps the only operations connected with textile work which had been taken up entirely by organized industries. Our most abundant and characteristic material evidences of domestic spinning are perhaps the spindle-whorls, usually of terra-cotta, which have been found in great numbers in various parts of the Græco-Roman world. Household weaving seems to be represented by the round bits of bone, pierced with one or more holes in the side, which have been regarded as fragments of flutes or as box-hinges, but, according to a recent hypothesis, were weights for keeping the threads taut in the weaver's loom, in much the same way as the wooden bobbins used by modern lace-makers or, indeed, the counter-weights on a modern loom.

Furniture and other articles of wood may likewise have been made at home, at least to some extent. But here we lack archæological evidence.

These household crafts were carried on by the women and slaves of the family.

In the towns, industrial labour usually took the form of small trades. Very often the same premises served for workshop and sale-room, the owner, assisted by one or two slaves or a few free workers, himself making the things which he sold. Other workshops, which manufactured for a private clientèle, employed slave labour and free labour equally. Pompeii and Ostia, with their narrow shops arranged in the front of the ground floor of the houses, on either side of the entrance door, give us an idea of these modest workshops, in which no middleman came between producer and consumer. The premises of the bronze-workers at Alesia[1] were probably like those of the blacksmiths and goldsmiths at Bibracte (Autun), described by Bulliot and Déchelette.[2] Of the potters' kilns found in many parts of the Roman world, some certainly belonged to big potteries, but others, to judge from their small size, very primitive arrangements, and situation in private houses in the middle of towns or villages, can only have belonged to the workshops of small manufacturers.[3] In the East, in the towns of Asia Minor, Syria, Phœnicia, and Egypt, the small craftsman doubtless worked under much the same conditions as his descendant in the *sukh* of the modern Levant. The only difference can be that slave labour, if it has not disappeared altogether, has at least diminished.

The organization of the crafts all over the Roman Empire reveals how strong the guild-spirit had become in the early centuries of our era. The professional associations, whose numbers and vitality are attested by many inscriptions, were chiefly composed of small employers and free workers. In the lists which Waltzing and Ciccotti have made of them,[4] all trades, or almost all, are represented. One even finds a very curious specialization, similar to that often noticed in the guilds of the Middle Ages. In addition to cobblers in the general sense (*sutores*), a distinction was made between *caligarii*, *crepidarii*, and *solearii*, that is, the makers of three

[1] *Pro Alesia*, 1st ser., 3rd year (1908-1909), pp. 436 *ff.*

[2] G. Bulliot, *Fouilles du mont Beuvray de* 1865 *à* 1897; J. Déchelette, *Fouilles du mont Beuvray de* 1897 *à* 1901.

[3] H. Thédenat, in **XVII**, *s.v.* "Fornax"; at Alesia, *Bull. Arch.*, 1914, p. 402; *cf. Pro Alesia*, vols. ix-x, p. 81. The Punic potter's kiln found at Carthage in 1922, with the wheel formed of two blocks of hard sandstone, doubtless gives an exact idea of what the Roman kilns afterwards were in Africa (*Bull. Arch.*, 1923, p. lxxiii).

[4] **CIV**, vol. ii, pp. 145 *ff.*; E. Ciccotti, [*Il Tramonto della schiavitù nel mondo antico*] *Le Déclin de l'esclavage antique*, translated by G. Platon, pp. 419 *ff.*

different kinds of footgear, *caligæ*, *crepidæ*, and *soleæ*. Among the potters (*figuli*), the statuette-makers (*sigillarii*) and tile-makers (*tegularii*) formed separate associations. In the textile industry, there were numbers of special guilds, for the fullers (*fullones*), the wool-workers (*lanarii*), the linen-workers (*linarii*), the cloak-makers (*sagarii*), the tailors or dressmakers (*vestiarii*), and the purple-dyers (*purpurarii*). It was the same in the metal industries, where there were *aurifices* or goldsmiths, *ærarii* or bronze-workers, *ferrarii* or blacksmiths, and *plumbarii* or lead-workers. Many more examples could be given.

We cannot here study the trade-guilds of the Roman world in detail, but we must at least pay attention to their general character. They were, first and foremost, associations of men who plied the same trade, as is proved by their very names. Some historians, however, by rash comparison with the Mediæval guilds and in an excessive desire to say exactly what they were, have ascribed to them an economic rôle which they do not appear to have played. " It has been held," Waltzing says, " that the Roman colleges had a similar purpose " (to that of the Mediæval guilds), " namely, the protection of the trade, the improvement and preservation of industrial processes, and even the training of apprentices. An attempt has even been made to show that they were companies formed for joint enterprises. I am of opinion that all these statements should be taken with caution."[1] The true object of these associations was to give the workers more power to defend their common interests and to secure for them, what they had long lacked, the consideration and esteem due to them for their services to the community.

In addition to this main purpose, the Roman guilds had a moral influence on their members, by bringing them together in a common worship, giving even the poorest the certainty of a decent burial, and coming to their aid in case of sickness or accident. " Religion, concern for their funerals, the wish to have more power to defend their interests and to rise above the mass of the proletariate, the desire to be friendly together and to make their hard life more pleasant—these were the various sources of that imperative need to associate which was so strong in the lower classes."[2]

[1] **CIV**, vol. i, p. 182.

[2] *Ibid.*, p. 333.

Although these small and medium-sized trades had a great place in the industrial economy of the Roman world, and although the development of the associations gave them great strength, they could not suffice for all the needs of the ancient world. There were also big workshops, large manufactures, one might almost say firms, corresponding to the big industry of our own times. This, indeed, was the only form in which certain industries could be conducted.

Many of these big workshops we know to have been organized on huge landed estates, in which industry was combined with agriculture and stock-breeding. In the villa at Martres-Tolosanes, south of Toulouse, a weaving-mill has been identified. Since the staff employed in this villa amounted to 200 or 300 persons, it is possible that the mill was set up solely to supply their needs. But there is no proof that the cloth made here was not sold outside, at least in part, thus figuring among the revenues of the estate.[1] Similar observations have been made in several villas in Britain, such as those at Darenth in Kent and Chedworth in Gloucestershire, where there were fullers' shops.[2] Mr. Rostovtzev believes that the Batavian and Frisian garments which had such a reputation all over the Roman West were made under the same conditions.[3] In Roman Belgium, "the villas," M. Cumont writes, "were busy hives, in which swarms of slaves or day-workers plied all the crafts needed for the working of the estate and sometimes for export."[4] Further on, with reference to the ceramic industry, he says: "The big villas became centres of production, in which a series of workshops employed a quantity of slaves and day-workers and supplied an extensive clientèle."[5] Even the metal industry took on this form at some places in Belgium; several villas possessed foundries or forges, chiefly for iron, for instance at Neufchâteau near Jemelle in the Province of Namur and Latinne and Louvignie in the Province of Liége.[6] The best-known of these villas, which were as much industrial as agricultural, is that at Anthée in the Province of Namur, which has yielded

[1] L. Joulin, in **XI**, 1st ser., vol. xi (1901), p. 287.
[2] **CI**, p. 308.
[3] **XCVIII**, pp. 166, 539 (n. 39).
[4] **XC**, p. 43.
[5] *Ibid.*, p. 65.
[6] *Ibid.*, pp. 38-39.

a furnace for smelting bronze and a great number of manufactured articles, small objects or pieces intended for furniture fittings, which speak of the skill and taste of the founders and craftsmen of the country of the Tungri and Eburones.[1] An inscription found near Thugga (Dugga) in Tunisia reveals the existence of a great private estate belonging to the Pullæni (*prædia Pullænorum*). Now, the signature *Pullæni* or *Pullænorum* is found on a fairly large number of terra-cotta lamps discovered in North Africa and Sardinia. It is therefore likely that one centre of this manufacture, doubtless the chief, the head factory, as we should say, was on the rural estate near Thugga.[2] An indication of a more general kind is furnished by a passage in the Digest (Book XXXIV, Tit. iv, Law 16, §11): Caracalla rules that if, on an estate, a *colonus* or slaves of the owner manufacture iron contrary to the law, without the knowledge of the owner, the latter shall not be held responsible.

So the owners of great landed estates did not hesitate to supplement the revenues which they obtained from agriculture and stock-breeding by the profits of various industries, weaving, pottery, metal-working. It is probable, however, that the workshops thus maintained supplied only local needs, or at the most those of the surrounding district.

In the Imperial epoch there were also industrial organizations with a far bigger scope, some private and others public and official. The chief potteries of Italy and Gaul, those of Arretium in Etruria, of La Graufesenque among the Ruteni, of Lezoux among the Arverni, of Rheinzabern among the Nemetes, and yet others, have the appearance of powerful industries, exporting their output to distant markets and holding almost a monopoly of such trade. The countless terra-cotta vessels signed by the Cn. Ateii of Arretium, by Libertus and Paternus of Lezoux, by Mommo of La Graufesenque, and by Cerialis and Cobnertus of Rheinzabern, to mention only the most abundant, bear witness to intense production. The lamps of Fortis, specimens of which have been found all over Italy and in Gaul, Britain, and Dalmatia, those of C. Junius Draco, which spread to Italy, Sicily,

[1] **XC**, pp. 75-76.
[2] J. Toutain, in **XVII**, *s.v.* "Lucerna," p. 1332.

Sardinia, Africa, and Narbonensis, and those of Strobilus, which are frequent not only in Italy but in Dalmatia and Pannonia, evidently came from potteries with a considerable plant. The Gallic potter Pistillus, whose terra-cotta statuettes have been found in every part of Gaul, and who was perhaps established in the country of the Ædui somewhere near Autun, gives one the impression of a big manufacturer. The master glassmaker Frontinus ran a considerable factory, "employing hundreds of workmen or clerks and possessing branches and separate workshops, each under a foreman, who was a slave or freedman of the owner."[1]

In metallurgy, apart from mining, we know of the Campanian P. Cipius Polybus, whose pots and pans made their way into the Rhenish provinces and even went beyond the limits of the Empire to the shores of the Baltic;[2] the Belgian Aucissa, whose bronze fibulas and pins were in great demand; and the Helvetian Gemellianus, who had a factory at Baden in Switzerland, and specialized in sheaths for gladiators' swords.[3]

We have no such explicit evidence for the textile industry. It is, however, probable that the materials, clothes, and carpets[4] which had a widespread reputation came from big mills, the owners of which jealously kept the secrets of their improved processes. Such, no doubt, were the *tunicæ russæ Canusinæ*, the *Tarantina*, the Maltese *othonia*, the *carbasa* of Tarraco, etc.

Establishments of this kind could only develop and flourish if their owners had considerable funds at their disposal. It has, therefore, been said that industry, in the last centuries of antiquity, had become, at least in part, capitalistic. But one must not exaggerate this character, nor, above all, regard the evolution as having been more general than it was. Household economy had not disappeared, as I have said, and small industries were still numerous in town and country. M. Jullian speaks rightly of an unassuming industrial middle class and even of an artisan democracy, "which were, in the best days of the Empire, the pacific and hard-working force in society,"[5] not only in Gaul, but pretty well everywhere.

[1] **XCVI**, vol. v, p. 310.
[2] *Ibid.*, pp. 303-304.
[3] *Ibid.*, p. 304.
[4] See above, pp. 296-7.
[5] **XCVI**, vol. v, p. 311.

The existence of municipal factories and State factories is proved by definite evidence for the third and fourth centuries only, especially the fourth. One may, however, legitimately assume that they already existed in the second and even the first century after Christ. We know, for example, that the legions themselves made the bricks and tiles which they needed for building their permanent camps and posts; that soldiers were often employed on the foundation of colonies, which involved public works and indeed a complete building-industry; and that sometimes quarries were worked either by soldiers or under army control.[1] In addition, it is likely that, as early as the first and second centuries, the Imperial government reserved the manufacture of arms and machines of war for State arsenals. What is certain, is that " there were, at the depôt of every legion, special workshops which seem to have been intended chiefly for repairing weapons and keeping them in good order, but, needless to say, were organized so as to be able to make them." [2]

So, in the first centuries of the Christian era, in the West as in the East, industry was active and prosperous, in every form, from the domestic work done round the hearth in every house to official manufactures organized in the workshops of the State. In all these forms, the industry of Imperial times employed both slave labour and free labour; there were slave workers as well as salaried workers in the towns, and salaried workers as well as slaves in the country, the free workers in the country being often combined in guilds which played a considerable part socially. For lack of exact evidence, we cannot determine the proportion of slaves to free workers or the importance of each element in the return of labour and the total of production.

[1] R. Cagnat, in **XVII**, *s.v.* " Legio," p. 1063.
[2] **LXXXIV**, p. 337; *cf.* pp. 172 *ff.*

CHAPTER IV

TRADE. EXCHANGES. ECONOMIC CIRCULATION. THE EQUIPMENT OF TRADE[1]

As we have seen in the last two chapters, agricultural, pastoral, and industrial production had developed enormously in the first centuries of the Empire, but at the same time it had, at least in part, become specialized. The result was that in many regions the exploitation of the soil or the manufacture of goods did not meet the needs of the consumers, while other regions produced more of some particular commodity or merchandise, metallurgical, textile, or ceramic, than it consumed. So there arose an economic circulation, in addition to which there was the transport of raw materials from their place of origin to the various places where they were used, such as the ingots of lead which travelled across Gaul to the Mediterranean ports.[2]

Moreover, Rome and the greater part of Italy had become a very powerful centre of attraction for many natural and manufactured products. Being dependent on the rest of the Empire, since their agriculture had declined and their industry had developed only in some places, the capital of the world and the Italian peninsula, with their large agglomerations of population, literally could not have lived without constant and abundant imports. Moreover, most of the inhabitants of Rome and the chief towns near it were not content with the necessities of daily life. Luxury had become a necessity to them; the taste for products of distant countries had spread beyond all bounds, not only among members of the Senatorial and Equestrian orders, but even among the freedmen, who had made fortunes, sometimes very large ones, by trade or banking. So goods of every origin, every

[1] **CVI**, pp. 18B-18C.

[2] M. Besnier, in **XII**, 1920, ii, pp. 211 *ff.*; 1921, i, pp. 36 *ff.*; 1921, ii, pp. 98 *ff.*

kind, and every value came pouring into Rome and the ports of Italy.

Under the influence of these economic, psychological, and moral conditions, and thanks to a remarkable development of ways of communication and means of transport, a commercial movement of hitherto unknown intensity set in from one end of the Roman world to the other, from the Scottish mountains to the fringes of Arabia and from the banks of the Danube and the coasts of the Euxine to the sands of the Sahara and the slopes of the Atlas. It went beyond the frontiers of the Empire; in addition to the internal movement there was an external traffic, which by the second century reached Central Asia, the Far East, Equatorial Africa, and the shores of the Baltic.

I

Articles of Commerce Inside and Outside the Empire

The articles of commerce within the Empire, which were transported not only from the various provinces to Rome and Italy but from one province to another, were of the most varied kinds. Foodstuffs consisted of the corn of Sicily, Africa, and Egypt; the oil of Spain and Africa; the wines, natural or manufactured, of Gaul, Dalmatia, Asia Minor, and Syria; the salted goods of Spain, the Euxine, the Cimmerian Bosphorus, and Egypt; and the pork of Gaul and Britain. The raw materials sent about the Empire in the greatest quantities were the flax of Egypt and Africa; the wool of Asia Minor and Africa; the hides of Illyricum and Asia Minor; the leather of Spain, Gaul, and Britain; the marble of Greece, Asia Minor, and Africa; the metals of Spain, Britain, Illyricum, Cyprus, and the country of the Chalybes in the depths of Asia Minor; and the timber of Africa and Syria. Many cloth-stuffs, sometimes embroidered, dyed purple, or woven with gold, came from the workshops of Spain, Gaul, Asia Minor, Syria, Phœnicia, and Egypt. Gaul and Asia Minor supplied the Empire with pottery and terra-cotta statuettes. Gaul, Phœnicia, and Egypt were renowned for their glass. Jewels made in the East were especially prized.

The perfumes which came out of the works of Asia Minor, Syria, Phœnicia, and Egypt were preferred to all others. Cyrenaica exported silphium, Egypt papyrus, and Pergamon parchment to all parts of the world. The forests of North Africa furnished wild beasts for the games of the circus and the combats of the amphitheatre. Asia Minor, Syria, Egypt, and Africa sold slaves.[1]

As we see from this list, West and East contributed about equally to the movement of trade, Africa, Spain, and Gaul in the West and Asia Minor, Syria, Phœnicia, and Egypt in the East.

These natural and artificial commodities, produced by the soil of the provinces or by the labour of their inhabitants, were not the only ones which fed the trade of the Roman world. Similar or different products came in from the distant lands of the North, East, and South—amber and slaves from the Baltic and Germany; corn, hides, furs, and precious stones from Scythia; slaves and wild beasts from the plateau of Iran, now held by the Parthians; alabaster, frankincense, myrrh, spices, and precious stones from Arabia; spices, perfumes, ivory, ebony, wild beasts, slaves, and precious stones from India; raw or manufactured silk from China; tortoise-shell and ivory from East Africa; and gold dust, hides, wild beasts, ivory and black slaves from Equatorial Africa by way of the Sahara.[2]

All this was almost wholly import trade. No doubt, countries like Germany and India occasionally bought natural products of the Mediterranean world or goods manufactured in the Empire from the merchants who visited them—wine, glassware, metal utensils, works of art, and rubbish. But what those traders chiefly gave in exchange for the wares for which they went to distant lands was money. That exportation of gold and silver coins had its disadvantages, and caused the Emperors some anxiety.[3] At least, what the facts and documents now known allow us to say with certainty, is that the Roman Empire, regarded as a whole, was, thanks to the variety of regions of which it was composed,

[1] *Cf.* the table drawn up by R. Cagnat and M. Besnier, in **XVII**, *s.v.* "Mercatura," p. 1778.

[2] *Cf.* the aforementioned table; also **XCVII**, pp. 293 *ff*.

[3] Cagnat and Besnier, *art. cit.*, p. 1774.

quite capable of sufficing to itself in its first centuries; it went abroad only for luxurious goods such as silk, spices, perfumes, and ivory. Thanks to the Roman Peace, the Mediterranean world had reached a really stable economic equilibrium.

II

Communications by Land and Sea. Harbours. Conditions of Traffic: Customs, *Octrois*, Tolls.

One of the causes which most effectively contributed to the creation and maintenance of that equilibrium for at least two hundred years was the establishment of a very highly developed system of communications, which was perfectly adapted to all military and commercial needs.

On land, there were roads and waterways. "In Imperial times, the Roman roads formed one vast continuous system, with the capital of the Empire at the centre. A man might travel by land all the way from Rome to the Pillars of Hercules, to the western tip of Brittany, to the mouth of the Rhine, to the mouth of the Danube, to Byzantion, to Athens. Beyond the *fretum Gallicum*, the roads of Britain carried on those of Gaul. Beyond the Bosphorus, the roads of Asia Minor continued those of Thrace, and ran on to Nineveh and Babylon in one direction and to the Nile in another. From the Nile to the Atlantic, there was the coast-road of North Africa, from which branches ran into the interior of Egypt, Proconsular Africa, Numidia, and the Mauretanias; at Alexandria, it connected with the route to Asia, and there were short, easy sea-passages from Carthage to Lilybæum and from Tingis to Gades, where the Italian and Spanish roads ended. In every region, a great number of secondary roads branched off from the main arteries in all directions and served important districts. Nowhere was there an impassable break or gap. This very complete and well-designed system made it possible to provide for the defence of the frontiers, ensured the good administration of the provinces, and facilitated travel and trade between the most distant countries."[1] We should note, too, that the defence of the frontiers and the

[1] M. Besnier, in **XVII**, *s.v.* "Via," p. 790.

good administration of the provinces did just as much for commercial prosperity as ease of travel and trade between the different regions of the Empire. The highest and wildest mountain-ranges no longer offered any serious obstacle to the movement of trade. Three important roads crossed the Pyrenees, and the chief passes of the Alps, from Mont Genèvre to that of Adelsberg between Triest and Laibach, had been engineered and were crossed by an active traffic. The Balkan Peninsula was traversed by roads which ran, in spite of the confused contours of the country, from the Adriatic and the Ionian Sea to the Euxine and the Ægean. The wild plateaus of central Asia Minor, the mountains of Armenia, the desert of Arabia Petræa, even the Sahara, had not prevented the creation of routes, which were in regular use, from the Iranian plateau to Anatolia, from the valley of the Euphrates to Syria, and from the oases of the Fezzan and Ghadames to the coast of Tripolitana. That work, M. Besnier very rightly says, " does great honour to the practical genius of the Romans," whatever inspiration or help their engineers may have got from roads or tracks already existing in some places before the Roman conquest.

In addition to the road-system, there were waterways in some countries. Strabo, at the very beginning of the Empire, had observed and pointed out the fortunate disposition of the rivers of Gaul.

> " The whole country is watered by rivers, some coming from the Alps and others from Cemmene (the Cevennes) and the Pyrenees, some falling into the Ocean and others into our sea. As a rule, they flow over plains or along hills which are not steep enough to prevent navigation. The rivers are, moreover, so happily placed in relation one to another that it is easy to convey goods from one sea to another; goods have to be taken a short journey overland, but it is quite easy and over flat ground, and they travel most of the way by the rivers, up one and down another." [1]
>
> Further on, the geographer lays stress on " the kind of symmetry of the country in respect of its rivers and of the two seas, Inner and Outer. For anyone who understands the situation will see that this is not the least advantage of the country, in that it facilitates the exchange of the necessities of life between all parts of the population and establishes a community of interests among them; especially today, when, being released from all fighting, they are keenly applying themselves to agriculture and adapting themselves to civilized ways." [2]

Strabo points to the waterway formed by the Rhone, Saône, and Doubs, and to that of the Seine, by which goods

[1] iv, 1, 2. [2] iv, 1, 14.

are carried to the country of the Lexovii (Lisieux) and Caleti (Seine Inférieure), to be conveyed thence to Britain. He also mentions the waterway of the Loire, and that of the Aude, which is connected with the Garonne by a portage between 700 and 800 stades in length (80-90 miles).[1]

In most countries with a Mediterranean climate, the torrential habits of the watercourses almost entirely prohibited river-navigation, but this was not so with the Rhine, Danube, Euphrates, and Nile. South of the Caucasus, too, trade made use of the Phasis, which flows into the Euxine, and the Cyros, which flows into the Caspian. In any case, the use of the rivers of the East as trade-routes was nothing new; even the Greeks had taken it on from the ancient empires.[2]

Active as traffic was on the roads and some waterways of the Empire at that time, it was chiefly by sea that the different regions of the Roman world exchanged their merchandise of all kinds. On the sea the routes are not traced so exactly and have not such material fixity as land-roads and rivers. But at least we are certain of the ends of them, which were the great trading ports. Even on the routes themselves there were points by which ships, by the nature of things, had to pass, such as the Bosphorus, the Hellespont, and the Strait of Messina. The main lines of marine trade in the Mediterranean under the Roman Empire ran east and west and from various points on the circumference to the centre, that is, to the chief ports of Italy. In the East, the most important points from which these lines started were, going southward, the ports of the northern and eastern Euxine, namely, Theodosia, Panticapæon, Phanagoria, Tanaïs, Dioscurias, and Trapezus; Ephesos in Asia Minor, of which Strabo tells us that, in spite of the silting up of the harbour, "it is the biggest commercial centre in Asia west of the Taurus;"[3] Laodiceia and the ancient Phœnician ports of Tyre and Sidon in Syria; and Alexandria in Egypt. On the north coast of Africa, overseas trade was concentrated in the ports of Tripolitana, especially Leptis Magna, where the chief caravan-routes of the Sahara ended, and Carthage, which had been revived by Cæsar and Augustus. In Western Europe, Gades in Spain and Narbo and Arelate (Narbonne and Arles) in Gaul maintained regular and frequent relations with

[1] iv, 1, 14. [2] See above, pp. 144 *ff*. [3] xiv, 1, 24.

the East and with Italy. At the head of the Adriatic, Aquileia had acquired great commercial importance since Rome had become mistress of the Danubian countries.

"Merchant vessels," Strabo says, "have only to go up the Natiso 60 stades at the most (about 7 miles) to get there. A market has been established there for the Illyrian tribes of the Ister. They come for goods brought by sea, such as wine, which they load on to waggons in wooden barrels, and oil; and they give slaves, cattle, and hides in exchange."[1]

In the middle of the Mediterranean, in addition to Corinth, which had recovered its marine and commercial activity since the founders of the Empire, breaking with the unfruitful policy of the Republican Senate, had restored it to life, the great trading ports were Puteoli and Ostia, where heavy vessels put in from the four quarters of the horizon, laden with goods of all sorts for Italy and the capital of the world. The supremacy of Puteoli, which had been undisputed under the later Republic, was somewhat diminished by the improvements effected in the harbour of Ostia by Claudius and Trajan, but it nevertheless continued to be, under the Empire, the resort of the ships of all countries, the real outer port of the capital. The prosperity of Ostia, thanks to those same improvements and still more to the proximity of Rome, increased wonderfully in the second and third centuries. Lastly, the port of Rome itself, the Emporium, on the left bank of the Tiber, not far from the end of the Aventine, received a great part of the cargoes brought from various parts of the Empire, which were transshipped at Ostia from the large sea-going vessels on to barges which could go up the river. "Rome," it has been said, "being turned into a seaport by the proximity of Ostia, became from an economic point of view the chief controlling centre of the world market."[2]

But the Mediterranean was not the only sea which played a part in the economic activity of the Roman world. That activity overflowed westwards and north-westwards on to the Atlantic and the seas of Northern Europe, and south-

[1] v, 1, 8. For the economic importance of Aquileia, see **XCVIII**, pp. 71 *ff.*, 157.

[2] **XCV**, p. 306; Cagnat and Besnier, in **XVII**, *s.v.* "Mercatura," p. 1783. For Puteoli, see C. Dubois, *Pouzzoles antique*, *passim*; for Ostia, **CIII**.

eastwards on to the Indian Ocean. In both directions, commerce had assumed ampler dimensions.

On the Atlantic, the busy ports were Gades in Spain and Bordeaux and Boulogne in Gaul. According to Strabo, Gades sent many vessels on to the Outer Sea (the Atlantic) as well as the Inner Sea.[1] Bordeaux was in commercial relations with the western coasts of Spain and with Britain.[2] The harbour of Boulogne was illuminated by a lighthouse two hundred feet high, and its wharves were heaped with goods going to or coming from Britain.[3]

At the other end of the ancient world, the ports of the east coast of Egypt, Myos Hormos and Berenice, served the lines of shipping which, regularly since Hippalos had discovered the monsoons, did trade with India. Other vessels were content to put in on the southern coasts of Arabia, where certain places carried on for the transit of goods from India and the Far East. Yet others made southwards along the coast of East Africa, called at Adulis, the port of the Axomites (in the bay of Massowah), turned Spice Cape, and advanced into Equatorial waters. From Myos Hormos and Berenice two roads ran, connecting the Red Sea with Coptos on the Nile.[4] It was by them that most cargoes reached the port of Alexandria, to be distributed all over the Mediterranean. The canal connecting the top of the Red Sea with the Nile seems to have played only a secondary part under the Roman Empire.

Land-roads, waterways, and shipping-lines on and outside the Mediterranean provided the trade of the world under the Roman Empire with an admirably co-ordinated system of regular and easy communications.

On all these ways, traffic was not absolutely free. Goods conveyed across the Empire had to pay a number of duties, which were classed together under the general name of *portorium*. " The *portorium* corresponded to three kinds of duties or taxes known among ourselves—customs duty, that is, a tax collected at the frontier of a province or state for the benefit of that state; *octroi*, that is, a tax levied on goods entering or leaving a town, for the benefit of that town; and

[1] iii, 5, 3. [2] **XCVI**, vol. vi, p. 379. [3] *Ibid.*, pp. 458 *ff.*
[4] Cagnat and Besnier, *art. cit.*, pp. 1781-82; **XXXIV**, *Prov.*, English vol. ii, p. 280.

toll, that is, a fee demanded for going through certain places, for example, across a bridge. The Romans never had words to differentiate these various forms of similar taxes. . . . The only distinction which the Romans made was to divide them into land *portoria* and sea *portoria*, according as the duty was collected on the land boundaries of the Empire, of a province, or of a city, or at a seaport."[1]

The whole Roman world was divided, from the point of view of customs, into ten districts, other than Italy. They were: Sicily, Spain, Gaul, Britain, Illyricum, the provinces of Asia, Bithynia-Pontos-Paphlagonia, Syria, Egypt, and the whole of North Africa. The rate of customs duty varied in different districts, from 2 per cent. in Spain to 5 per cent. in Sicily, Africa, and Illyricum, so far as our present evidence allows us to say. The *octroi* about which we have most information is that of Palmyra. Of the tolls, we can only say that they existed, having no details about their organization and distribution. These various taxes, which were levied on goods all through the journey, considerably increased their price. Pliny the Elder tells us that goods from India were sold, when they reached their destination, at a hundred times their original value. The collection of the *portorium* in its various forms was of great importance to the finances of the State.[2]

III

The Organization and Equipment of Trade

The forms and methods of commercial activity in the Roman world were as varied as they are among modern peoples.

In every one of the innumerable towns of the Empire, in addition to the private shops of which we can obtain a fairly accurate notion from the ruins of Pompeii,[3] there was a public market or *macellum*, the stalls of which were usually arranged round a rectangular space, sometimes under porticoes.[4] Besides these permanent, daily markets, there were excep-

1 R. Cagnat, in **XVII**, *s.v.* "Portorium," p. 586.
2 *Ibid.*, pp. 587 *ff.*; **LXXXV, LXXXVI**, *passim*.
3 **LXIX**, pp. 190 *ff.*
4 H. Thédenat, in **XVII**, *s.v.* "Macellum," p. 1457.

tional markets (*nundinæ*), which were held at fixed intervals, usually twice a month, either in the towns or in the country.[1] We have the text of the Senatus Consultum which authorized a big African landowner, Lucilius Africanus, to hold such a market on his own estate every month on the fourth day before the Nones and the twelfth day before the Kalends; the document is dated A.D. 138.[2] These were only organs of local, or at the most regional, trade.

There were commercial centres in the Roman Empire of a very different kind, which one may call international, or at least interregional. These were chiefly the great ports of the Mediterranean, either in the middle, that is, in Italy, or on the circumference, from the Pillars of Hercules to the end of the Euxine and the coast of Phœnicia. These ports were animated by a considerable movement of trade. Not only did each one of them export the goods of its own particular district and import others for consumption on the spot or in the immediate neighbourhood; they were also nearly all of them at the ends of commercial highways, most of which came from long distances and brought the goods which were produced outside the Roman world. This was the case with the ports of the Cimmerian Bosphorus; with Dioscurias and Trapezus; with Ephesos, at the end of the great road from Central Asia; with the ports of Phœnicia and Palestine, which received from their back country everything going westward from the valley of the Euphrates and Arabia; with Alexandria, which was still the point by which most goods from India and East Africa entered the Mediterranean world; with Leptis Magna and the ports of Tripolitana, where the caravan-roads of the Sahara came down to the sea; with Carthage, whose wharves were piled with North African goods of all kinds, bound for Italy; with Narbo and Arelate, where the raw materials and manufactures of Britain, Gaul, and the Rhenish border which had come over Gaul by various routes were shipped; and, lastly, with Aquileia, the port of Illyricum and the Danubian provinces.

In addition to these ports, whose importance and activity are easily explained, there were inland cities which were able to play a similar part, thanks to their geographical position.

[1] M. Besnier, *ibid.*, *s.v.* "Nundinæ," p. 122.
[2] *Ibid.*; *cf.* V, vol. viii, 11451.

In Gaul there was Lyons (Lugdunum). Standing at the crossing of the great roads which ran through Aquitania and Lugdunensis and yet further north and north-east through Belgium and Upper and Lower Germany, standing, too, on the confluence of the Saône and the Rhone, crowded with cargo-boats, standing, lastly, on some of the roads which came over the Alps, Lyons, with its motley population of Gauls and Britons, Africans and Syrians, saw the whole world flocking within its walls to buy and sell. The business quarter (now Fourvière) "was like Athens or the Piræus."[1] No city in the West was of such economic importance.

In the East, several cities, while not rivalling the Mediterranean ports, had quite a special position on the great trade-routes. According to Strabo, Apameia Cibotos, in the south of Phrygia, was one of the great markets (ἐμπόρια) of Asia, "next in importance to Ephesos, which is the general entrepôt of goods from Italy and Greece."[2] Apameia was, moreover, one of the principal stations on the road by which goods came to Ephesos from Mesopotamia and Iran. East of Syria, almost half-way between the Mediterranean coast and the valley of the Euphrates, Palmyra had been, since the early centuries of the Empire, a very busy centre of the caravan-trade. The inscriptions discovered in its ruins tell us that the leaders of these caravans and great merchants were among the most respected burgesses.[3] Further south, on the eastern and southern borders of Palestine, Bostra and Petra, as middlemen between the East and the Mediterranean, enjoyed, from the second century onwards, "a fabulous prosperity,"[4] due to trade in the products of southern Arabia and Babylonia. In Egypt, Coptos on the Nile was the great transit centre. There the two roads from the ports of Berenice and Myos Hormos ended and goods from various regions on the Indian Ocean were re-embarked on the Nile for Alexandria.

The trading community naturally varied with the character of the trade and the amount of business. In each town the merchants combined in guilds, like the workers and manufacturers. In some of the guilds now known to us the members were engaged both in trade and in industry,

[1] **XCVI**, vol. vi, pp. 515 *ff*.

[2] xii, 8, 15.

[3] **XXXIV**, *Prov.*, English vol. ii, pp. 98-99.

[4] *Ibid.*, p. 155.

such as the bakers, goldsmiths, drapers, potters, and shoemakers, but others were composed entirely of merchants. We may mention, for example, the tavern-keepers (*caupones*), druggists (*farmacopolæ publici*), dealers on the Forum (*Forenses*), poultry-dealers (*gallinari*), butchers (*lani, laniones*), wood-merchants (*lignari universi*), merchant tailors (*negotiatores artis vestiariæ et lintiariæ*), sellers of preserved vegetables (*negotiatores salsari leguminari*), wine merchants (*negotiatores vinarii*), greengrocers (*holitores*), fruit-dealers (*pomari*), retailers (*propolæ*), and perfumers (*unguentari*).[1] M. Waltzing justly observes that, according to the inscriptions now known, the distribution of these guilds does not seem to have been the same in the various provinces. They were very numerous in Rome, at Ostia, in Cisalpine Gaul, at Pompeii, in Narbonensis, and at Lyons, fairly widespread on the Rhine and Danube and in Dacia, extremely rare in Greece and the East, and almost unknown in Africa. His remark applies to all the colleges taken together, industrial and commercial, but it is equally true of either category separately. The fact remains that there was combination among the merchants, who were mere middlemen, just as much as among the manufacturers, who were real producers. The idea of association was one of the original features of the economic life of Rome.

The boatmen who carried goods on lakes and rivers and the shipowners of the seaports likewise combined in colleges. The former, who are generally called *nautæ*, but sometimes also *navicularii amnici, scapharii, lintearii, ratiarii,* and *utricularii*, took their title either from the river on which they plied or from the town or tribe which was their centre of action. The inscriptions speak of the *nautæ* of the Rhone and the Saône, the *nautæ* of the Loire, the *nautæ* of the Moselle, etc., or else of the *nautæ* of the Parisii, the *nautæ* of Comum on the Lake of Como and of Brixia (Brescia) on the Lake of Garda, the *scapharii* of Seville, etc. Such guilds existed in Spain, Gaul, the Rhenish and Danubian provinces, north Italy, and even Egypt, on the Nile.[2] The existence of these guilds, some of which seem to have been powerful, like that of the *nautæ* of the Rhone and Saône at Lyons, shows the

[1] **CIV**, vol. ii, pp. 145 *ff*.
[2] *Ibid*., pp. 29 *ff*.; M. Besnier, in **XVII**, *s.v.* "Navicularii," pp. 21 *ff*.

importance of inland navigation to the movement of trade in the Roman Empire.

The shipowners (*navicularii, navicularii marini, ναύκληροι*), whose vessels carried the trade of the whole Mediterranean, likewise combined in colleges. In the time of the Antonines, these colleges were free, private associations. Later they became official institutions, and the shipowners were practically turned into State officials. We hear of colleges of shipowners confined to certain ports—Ostia and Tarracina in Italy, Arelate and Narbo in Gaul, Tomi in Lower Mœsia, Smyrna and Ephesos in Asia Minor, Arados in Phœnicia, Alexandria in Egypt. Others comprised the shipowners of a whole district or of all the ports of a particular sea; several documents mention the Shipowners of Sardinia, the Shipowners of Africa, the Shipowners of the Spains, the Shipowners of the Adriatic.[1] Since the whole Empire was distributed round the Mediterranean, these shipowners were one of the most important factors in the great movement of trade which traversed every part of that sea.

Of the part played by these shipowners in the traffic which brought the food and goods needed by Rome and Italy to the ports of Rome, Ostia, and Puteoli, we have ample and explicit evidence. There are, first, the inscriptions painted on many fragments of amphoras found on the Monte Testaccio, a mound 115 feet high and half a mile round, which is entirely composed of potsherds. "These amphoras, which were landed at the Emporium with their contents, formed the indispensable material of the cargo-boats. They were exposed to various accidents as a result of the handling to which they were subjected at the port. By having his name painted on these vessels, the shipowner gave his agents a means for recognizing them and avoiding disputes."[2] The scholars who have studied these curious documents, Father Bruzza, Dressel, Reifferscheid, Otto Hirschfeld, H. de Villefosse, have observed that most of the shipowners whose names have survived were Spaniards, Africans, and Gallo-Romans. Other inscriptions tell us that many Orientals, from Asia Minor, Syria, Phœnicia, and Egypt, lived in Rome, Ostia,

[1] CIV, vol. ii, pp. 34 *ff.*; Besnier, *art. cit.*, pp. 23 *ff.*

[2] H. de Villefosse, "Deux Armateurs narbonnais," in *Mém. de la Soc. des Antiquaires de France*, vol. lxxiv (1914), pp. 153 *ff.*, 167.

and Puteoli under the Empire. Among these Orientals some must certainly have been the representatives in Italy of Eastern shipping agencies or mercantile houses.

The *stationes municipiorum* mentioned by Pliny the Elder[1] and Suetonius[2] were doubtless commercial offices in Rome, like the *stationes* of shipowners of a number of cities at Ostia[3] and that of the people of Tyre at Puteoli.[4] One hesitated over the meaning of the word *statio* before the discoveries at Ostia were made.[5] The expression *statio Sabratensium*, which is to be seen on the mosaic pavement of one of the *scholæ* near the theatre, proves that this word must be understood to mean the office owned in common by the shipowners of Sabrata in Africa. That these *stationes* at Rome, Ostia, and Puteoli[6] should have taken part in religious ceremonies, as appears from the inscription at Puteoli, or should have built monuments in honour of their native city is not at all unexpected in an organ of collective life in antiquity. The purpose and economic function of the *stationes* furthermore explain why in Rome they stand round about the Forum, and why several of them occupied premises in the house which was owned by Salvidienus Orfitus in Nero's reign. The site and arrangement are far more suitable to commercial agencies than to a kind of municipal embassies. Whatever may be the truth regarding the *stationes* of Rome and Puteoli, the excavations of Ostia have told us beyond all possible doubt that there were in that port agencies or offices of shipowners of several Mediterranean ports. Twenty-five of them occupied premises situated along the two long sides of a vast *area* near the theatre. Unfortunately, not all of their inscriptions have been preserved; only seven survive. Six mention towns in Africa, including Carthage, Hippo Diarrhytos (Bizerta), and Sabrata (in Tripolitana); the seventh bears the name of Turris, perhaps Turris Libisonis in Sardinia (Porto Torres). Another inscription at Ostia testifies to the presence in that port of the owners of the merchant fleet of Alexandria.[7]

From the information supplied by all these documents,

[1] *N.H.*, xvi, 44 (86). [2] *Nero*, 37.
[3] **CIII**, pp. 81 *ff*. [4] **IV**, vol. i, 421; *cf*. 132.
[5] Platner, *Topography and Monuments of Ancient Rome*, p. 176.
[6] **IV**, vol. i, nos. 111, 131-132.
[7] *Ibid.*, no. 392.

one may conclude that there was a similar organization in all the principal ports of the Empire, at least in those where there was much traffic. Thus, from one end of the Empire to the other, commerce was, as it were, unified. Natural commodities, raw materials, and manufactured goods travelled about the Mediterranean in every direction; they were carried by whole merchant navies, and associations of shipowners, organized in the busier ports, doubtless played the part of the modern shipping company.

Commerce on this scale needed convenient methods of exchange. For a long time coin had been the current instrument of sales and purchases in most Mediterranean countries, but each of those countries had its own coinage. In the East, it is true, in spite of political and territorial divisions, the circulation of a common coinage, or of equivalent coinages, had been successfully organized;[1] but in the West, before their annexation to the Roman world, Carthage and the Punic cities, the Iberian tribes, and the peoples of Gaul, every large group of the population, had struck different moneys. The issue of a coinage being, in theory at least, a consequence and a privilege of sovereignty, Augustus, after the establishment of the Empire, endeavoured to introduce the Roman monetary system into all the countries subject to Rome. Only the Emperor had the right to issue gold coins, and in theory he also had the sole right of issuing silver ones, but in practice he authorized certain cities in the East to continue striking them—Antioch, Alexandria, Rhodes, whose drachmas and tetradrachms corresponded fairly easily with the Roman denarius, and several cities in Asia Minor, whose *cistophori* were accepted by Roman traders.[2] He allowed the Senate to keep the right of issuing bronze coins, the chief of which were the sesterce and the *as*; the letters S C[3] borne by these coins bear witness to this minting by the Senate. In addition, the cities of the East were allowed to strike bronze coins, provided that they bore the Imperial effigy;[4] Athens alone seems to have been exempted from this obligation. At the beginning of the Empire, several towns in the West, colonies or *municipia*, Gades, Nemausus, and

[1] See above, pp. 166 *ff*.
[2] **XXI**, vol. x, pp. 44 *ff*.
[3] *S*(*enatus*) *C*(*onsulto*).
[4] F. Lenormant and Babelon, in **XVII**, *s.v.* "Moneta," p. 1975.

Lyons among them, had been granted the privilege of minting money, even silver. But it was a short-lived concession; before the fall of the Julio-Claudian line, colonial money, even of bronze, had wholly disappeared from the West.[1] In any case, the minting of bronze was of no importance for big commercial dealings or those with other regions.

These dealings were, moreover, settled less by payments in hard cash than by banking operations, which were conducted in Rome by the *argentarii*, whose shops (*tabernæ argentariæ*) were near the Forum or in the adjoining quarters, and by the *nummularii*, who went in more especially for the testing and, if need arose, exchange of money.[2] Both classes had to observe certain rules laid down by the law; they were subject to the control of the State. Outside Rome, one may assume that there were likewise *argentarii* and *nummularii* in all commercial centres of any importance.[3] In the East the banking system had long been in force.[4]

This organization of coinage and banking, coupled with the diffusion of Roman weights and measures, which, if not obligatory, was more or less inevitable, especially in the West, encouraged trade on a big scale and diminished the grave disadvantages which might have ensued from the very size of the Empire and the variety and multiplicity of pre-existing economic institutions.

It was in the second century of the Christian era, under the Antonines, that the economic life of antiquity knew its greatest days. By the extent and wealth of the territory exploited, by the abundance of the raw materials furnished by the soil of the provinces and imports from many foreign lands, by the development, activity, and specialization of numerous industries, by the expansion and organization of commercial relations, the Roman Empire for two hundred years enjoyed a prosperity which a few districts of limited size, or some cities which were specially favoured at certain times—Tyre and Sidon, Athens and Miletos, Alexandria,

[1] *Ibid.*, pp. 1976 *ff.*
[2] **XXI**, vol. x, pp. 80 *ff.*; G. Humbert, in **XVII**, *s.v.* "Argentarii," p. 407.
[3] See especially, for Gaul, **XCVI**, vol. v pp. 347 *ff.*
[4] See above, p. 168.

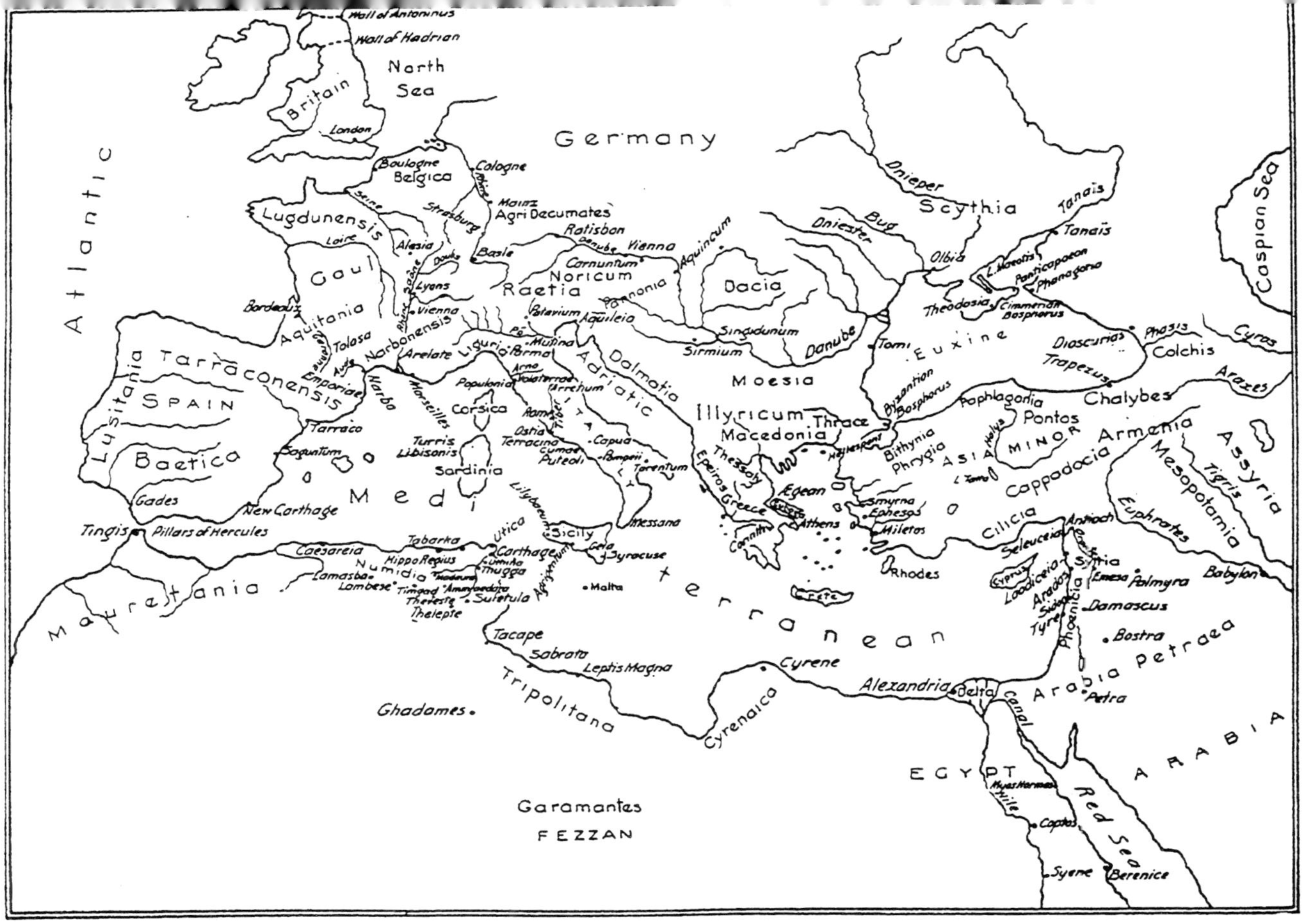

MAP VI. THE ROMAN EMPIRE.

[face p. 320.

Rhodes, Carthage—may have known in the past, but which had never extended over such an immense geographical area.

But decline came fast, and then downfall. The causes of the catastrophe were many. A study of those causes will serve as a conclusion to the economic history of the ancient world.

CONCLUSION

THE ECONOMIC DOWNFALL OF THE ANCIENT WORLD[1]

In the second century after Christ, under the dynasty of the Antonines, the economic life of the ancient world, after a long evolution starting from the agricultural, industrial, and commercial organization of Homeric society, had come to embrace, in all their extension and variety, the whole of the regions forming what is commonly called "the world known to the ancients." Even beyond the limits of that world, it had entered into almost regular relations with the north of Europe, Western and Central Asia and the Far East, and Eastern and Central Africa. It knew no barrier to its expansion but the then impassable immensity of the Atlantic Ocean.[2]

We have seen the causes of that prosperity—the security of the frontiers of the Empire, order and peace inside them, the political unity and administrative elasticity of the Roman State, an intelligent development of public works, and the spread of urban and settled life in almost all the provinces. No doubt, even in the early days of the Empire, the economic picture had its dark patches. Greece was in a deplorable condition of distress. In many parts of Italy agriculture was floundering in difficulties which were almost incurable, and it was in vain that Nerva and Trajan tried to save it by loans of money to landowners, the interest being devoted to the fund for the *pueri alimentarii*. Even before the end of the second century, symptoms of decay were appearing.[3]

Things grew rapidly worse in the third century. Everything conspired to the same effect. On every side, the frontiers of the Empire were forced, either by barbarian hordes, as on the Rhine and Danube, or by armies greedy for plunder, as on the Euphrates and in Syria and Asia Minor. The Franks, the Alamanni, the Goths, the Scythians, the Parthians, and later the Persians overran, looted, and laid

[1] **XCII, CII.** [2] See above, pp. 251 *ff.* [3] **XCV**, p. 337.

waste some of the richest provinces of the Empire. In Egypt, the Blemmyes, a race of Ethiopian marauders, made their way down the Nile as far as Coptos and Ptolemaïs. The most frightful of these inroads, the one which can give us a notion of the ruin which they left behind them, was the invasion of the Franks all through Gaul and into Spain in 257 and 275. The memory of it was still lively in the fifth century, when Paulus Orosius described how "in the midst of the ruins of great cities miserable dwellings stand, which bear witness to past calamities and have preserved for us the names of other days."[1] M. Jullian, in estimating the results of that invasion, has recently written: "The Germanic bands overran all Gaul from the Rhine to the Pyrenees. It was the end of a world; never again would our country see such calamity. All our cities were destroyed, the happy, amiable cities built by Rome, shining with marble and gay with statues, where for three hundred years no sound had been heard but that of crowds of men at work." The centre of the Empire, Italy, was not secure from these ravages; more than once the Alamanni appeared south of the Alps.[2]

On sea as on land, the barbarians did their disastrous work. Over the Euxine the Scythians came to loot the cities of Asia Minor, and sailed on to Crete and Cyprus. The Goths scoured the Eastern Mediterranean. In 280 a piratical band of Franks sailed right across the Empire from the Euxine to the Pillars of Hercules and made their way along the coasts of Iberia and Gaul to the mouths of the Rhine. So it was not only the outer circumference that the barbarians attacked and ravaged; no part of the Roman world was safe.

To invasions was added anarchy at home. The wars of the pretenders to the Empire, which went on almost without interruption for a century, sowed ruin everywhere. In 195, after his victory over Pescennius Niger, Septimius Severus ravaged Antioch, and two years later Lyons underwent the same fate because it had supported Clodius Albinus. In 238, when the province of Africa proclaimed Gordian Emperor, Capelianus, the Legate of Numidia, who remained loyal to Maximin, put the richest parts of North Africa to fire and the sword; cities, villages, and temples were given over to a brutal and greedy soldiery; the victims were beyond

[1] vii, 22.

[2] *Cf.* **LXXII**, pp. 8 *ff.*, 55 *ff.*

numbering. These examples give one a notion of the awful consequences produced by the military anarchy.

Sick of this treatment, the inhabitants of many provinces revolted. Riots and insurrections broke out all over the Empire. Brigandage raged in Italy and Sicily no less than in Africa and Asia. The rising of the Bagaudæ in Gaul was a perfect Jacquerie, which went on for years.[1]

Lastly, even the unity of the Roman world was destroyed for nearly twenty years by the attempts of Postumus and his successors in the West to found a Gallo-Roman Empire and by those of Odenathus and Zenobia in the East to found a Palmyrene Empire extending from Egypt to the Euphrates and Asia Minor.

The moral unity of Roman society was equally damaged by the persecution of the Christian religion in the third century.

" This accumulation of miseries, which were reinforced by terrible natural catastrophes, plague and earthquakes, spared none of the regions of the Empire, and, as was to be expected, produced the most disastrous effects in the economic domain. Shortage of production and impossibility of movement . . . money-shortage and high cost of living . . . depopulation and general ruin—the whole economic fabric of the State was cracking and seemed likely to break up at any moment."[2] The countryside was abandoned by such peasants as had escaped massacre, who took refuge behind the walls of the towns. More and more land fell to waste. In the towns, which had undergone repeated inroads, the trades were at their last gasp. The dangers attending travel by land and sea dealt commerce on a big scale a blow which was almost fatal. The wealthy and middle classes—the landowners, manufacturers, and merchants—were hardest hit, and it was to them that economic prosperity really owed its being. Taxation — land - tax, indirect taxes, *portorium*, etc. — no longer provided the resources which the treasury needed. Public wealth and private wealth, both struck at their sources, were vanishing.

No doubt, energetic efforts were made to cure the decay

[1] G. Bloch, in E. Lavisse, *Histoire de France*, vol. i, pt. ii, pp. 266, 321.
[2] **XCV**, p .341.

and to avert collapse. At the end of the third century and at the beginning of the fourth, the unity of the Empire was restored, in a new form, by the Illyrian Emperors, and then by Diocletian and Constantine. This was not effected without civil wars and violent struggles; it seemed to be an accomplished fact under Constantine. The monarchy of Oriental type, introduced by Diocletian and organized by Constantine, seemed to give a real authority to the Imperial government for a time. A vigorous counter-offensive against the barbarians, which went on to the end of the fourth century under Julian and Theodosius, cleared the frontiers of the Empire for a time, although certain outposts, such as Britain, the Agri Decumates, and Dacia, had to be given up. The Edict of Mediolanum was issued with a view to putting an end to religious wars. By these measures the government wisely attacked the very causes of the evil—anarchy, insecurity, and discord at home. Unfortunately, these efforts did not succeed. Neither the internal unity of the Roman world, nor the defence of the provinces against the barbarians, nor moral peace and concord were restored permanently. Nothing could prevent the *Romani corporis dilaceratio* which Orosius describes in such vigorous terms in the passage quoted above.

Lastly, and chiefly, the Imperial government imagined that it could overcome the economic crisis through which the Roman world was passing by means of remedies which proved worse than the disease. To prevent the desertion of the countryside, it endeavoured to pin the cultivators to the land; it therefore issued laws by which the *colonus* was tied to the estate on which he worked. If the *colonus* was not a slave, " he was like one in many respects. If a slave could not leave his master, the *colonus* could not leave the estate. He was bound to the soil as tightly as the slave to his master. He belonged, in a way, to the land ' for which he was born.' He had ' to serve it in perpetuity.' He was ' a human person due and subject to the soil.' . . . It is a significant thing that the lawgiver applies to the *colonus* terms which seem to suit only a slave."[1]

To prevent landowners who were members of municipal Curiæ from abandoning their land and evading the now heavy burdens to which they were subject on account of their

[1] XXXVII, p. 106.

financial responsibility to the State, the Imperial government issued many edicts which forcibly bound the Curiales to their condition. The mere fact that so many were issued show how ineffective they were.[1]

It was for the like reasons that the State turned the guilds of craftsmen and merchants into official institutions, having previously exercised only a mild control, a legitimate supervision, over them. The disasters which had reduced many towns of the Empire to penury had dealt a terrible blow to economic activity, industry and trade alike. All the same "corn must be carried, bread baked, lime burned, fires extinguished, and so on. The State entrusted each of these tasks to a definite class of citizens. . . . They were entrusted to the class of craftsmen or merchants whose profession made them specially suited for the work, and most of those craftsmen or merchants belonged to guilds, which gradually became official institutions, wheels in the administrative machine."[2] Waltzing compares the position of these guilds, thus enrolled in the government service, to that of the Curiales. "Their service was practically regarded as a *munus publicum*; consecrated by usage, it was imposed collectively on the body of members responsible for it, just as the Curiales were responsible for the collection of taxes, and in the fourth century *corporati* and *collegiati* vainly tried to escape it."[3] One of the most typical examples of this process, by which the guilds were taken over by the State, is that of the *navicularii*. These shipowners, who originally formed free companies and shipped goods of all kinds on their own account or for rich merchants as well as for the State, were all transferred to the public service under the Late Empire; the goods which they carried were government cargoes. "They were charged with the shipping of public supplies;" their work was done under the supervision of officials.[4] After surveying the various corporations, all of which were more or less victims of this principle, Waltzing concludes: "This system of labour was based on compulsion. The hand of the State was everywhere, its tyranny was everywhere. Everywhere workers were enrolled or retained by force. Nowhere could one find

[1] G. Lacour-Gayet, in **XVII**, *s.v.* "Curialis," p. 1632.
[2] **CIV**, vol. ii, pp. 16 *ff.*
[3] *Ibid.*, p. 17.
[4] *Ibid.*, pp. 53 *ff.*

private initiative or free labour. Now, force has never favoured the productivity of labour; industrial progress can only result from liberty. . . . Compulsion destroys individual energy; it disgusts us with a profession for which we are not made, which does not pay us, to which no interest attaches us. Work was bound to slacken, productivity was bound to decline, and trade was bound to fall off. It was a necessary consequence of the meticulous, tyrannical regulations which did away with all freedom."[1]

The Imperial government went further. The cost of living had mounted steadily in the third century. To bring it down, Diocletian published his famous Edict fixing maximum prices in 301. " After a long explanation of motives, in which the Emperor sets forth, not without eloquence, the considerations of public utility which have caused the State to intervene, the document comes to the remedy needed, the establishment of a maximum price for all commodities, not to be exceeded in any case. That general principle is followed by a tariff of prices, all given in denarii. Foodstuffs, raw materials, manufactured goods, salaries of the liberal professions, and wages of workmen are all there, in a complete and detailed list. . . . Penalties are necessary, and they are as severe as can be: death for a merchant who contravenes the clauses of the Edict by selling above the legal price; death for the buyer who abets him; death for those who buy up and hold illegal stocks."[2] The result of the measure was that all prices at once rose, and the strict execution of the penalties was of no effect. Very soon the attempt to enforce the Edict had to be abandoned. Julian tried to revive the experiment in 362, but failed as Diocletian had done.

Lastly, the Emperors tried to remedy the economic crisis by reorganizing finance and reforming the coinage. The reorganization and the reform doubtless did away with some of the abuses arising from the anarchy of the third century. But the results obtained were not more than somewhat ineffective palliatives. The mere weight of the taxes in kind and coin by which all classes of society were burdened was bound to prevent an economic revival of any value. As for the coinage, it is true that some improvements were made in minting and issue, and the new monetary system gradually instituted by Aurelian, by Diocletian and Constantine, and

[1] **CIV**, vol. ii, pp. 483 *ff.*   [2] **XCV**, pp. 350 *ff.*

by Julian represented "an enormous advance compared with the anarchy of the third century"; but it could not prevent the steady depreciation of silver and bronze.[1]

So, at the end of the fourth century and the beginning of the fifth, when the barbarians, having broken down the last ramparts which defended the Roman world, poured in south of the Danube and west of the Rhine, the economic life of the ancient world had not recovered from the disorder into which it had been thrown by the crisis of the third century. In the provinces, the invaders might still find wealth to loot, huge estates and sumptuous villas to occupy, and cities full of monuments and works of art to sack, but these were only the very impoverished remains of a prosperity which had steadily declined for two hundred years, for lack of peace, security, and liberty. The conditions required for the expansion and productivity of human labour in all its forms had disappeared at the end of the second century. They had not been restored since. The Imperial government had neither succeeded in removing the causes of the evil nor seen that excessive and despotic interference by the State was powerless to prevent the consequences.

From its beginnings, which were marked by activity chiefly of a domestic kind, and its end, when it was entirely subjected to the tyranny of the public authorities, the economic life of the ancient world passed in the course of centuries through an evolution which is most instructive. Its first steps forward were due to the passion of the Greeks for liberty and their love of initiative and adventure. By conquering Asia and by organizing his conquests in a spirit of tolerance and assimilation, Alexander opened a marvellous field of action to it. After Alexander, Rome, by uniting the whole Mediterranean, by diffusing the classical civilization all over the West and part of the centre of Europe, by the elasticity and liberality of her policy towards the conquered peoples, and by the radiation of her prestige beyond the limits of the Empire, gave it at least two centuries of really remarkable brilliance and expansion. It only began to decline when, under the influence of dangers at home and outside, the Imperial government believed that the public authority was able to direct the labour of men and could substitute its often depressing influence for the fruitful inspiration of liberty.

[1] **XCV**, pp. 353 *ff*.

BIBLIOGRAPHY

This bibliography may perhaps seem short. I have only mentioned such works and collections as are essential. I have not considered it necessary to give many references to secondary works dealing with matters of detail. The names of such works will be found in the big general works which I have frequently used, such, for example, as Paul Guiraud's *La Propriété foncière en Grèce* and *La Main-d'œuvre industrielle dans l'ancienne Grèce*, Droysen's *Histoire dè l'Hellénisme*, Bouché-Leclercq's *Histoire des Lagides*, J. Déchelette's *Manuel d'archéologie préhistorique*, S. Gsell's *Histoire de l'Afrique du Nord*, Camille Jullian's *Histoire de la Gaule*, R. Cagnat and V. Chapot's *Manuel d'archéologie romaine*, Rostovtzev's *Social and Economic History of the Roman Empire*.

[Certain English and other editions are added to this bibliography in square brackets; the footnotes, however, do not refer to the pages of these editions unless it is so stated.—*Trs.*]

I.—GENERAL COLLECTIONS OF INSCRIPTIONS AND COINS

Boeckh (A.), and Frank, *Corpus Inscriptionum Græcarum*, Berlin, 1828-77. I

— *Inscriptiones Græcæ editæ consilio et auctoritate Academiæ Regiæ Borussicæ*, Berlin, 1873, etc. . . II

Dareste (R.), Haussoullier (B.), and Reinach (T.), *Recueil des inscriptions juridiques grecques*, Paris, 1891-1905 III

Inscriptiones Græcæ ad Res Romanas pertinentes, Paris, 1901, etc. IV

Corpus Inscriptionum Latinarum, Berlin, 1863, etc. . V

Head (Barclay V.), *Historia Nummorum*, 2nd ed., Oxford, 1911 VI

Babelon (Ernest), *Traité des monnaies grecques et romaines*, Paris, 1901, etc. VII

II.—PERIODICALS

Bulletin de Correspondance hellénique VIII

Klio IX

Mélanges d'archéologie et d'histoire (French School at Rome) X

Mémoires présentés à l'Académie des Inscriptions et belles-lettres XI

Revue Archéologique XII

Revue des Études anciennes XIII
Revue des Études grecques XIV
Revue Historique XV
Revue and *Nouvelle Revue historique de Droit français et étranger* XVI

III.—ENCYCLOPÆDIAS. HANDBOOKS. GENERAL WORKS

DAREMBERG (Charles), SAGLIO (E.), POTTIER (E.), and LAFAYE (G.), *Dictionnaire des Antiquités grecques et romaines*, Paris, 1877-1919 XVII
PAULY and WISSOWA, *Realencyclopædie der klassischen Altertumswissenschaft*, Stuttgart, 1894, etc. . . XVIII
DÉCHELETTE (J.), *Manuel d'archéologie préhistorique*, etc., Paris, 1908, etc. XIX
CAGNAT (René) and CHAPOT (V.), *Manuel d'archéologie romaine*, Paris, 1917-20 XX
MOMMSEN (T.), and MARQUARDT (J.), [*Handbuch der römischen Alterthümer*, 1st-3rd eds., Leipzig, 1871-88] *Manuel des antiquités romaines*, French trans., Paris, 1887, etc. XXI
Mélanges Cagnat, Paris, 1922 XXII
GUIRAUD (P.), *Lectures historiques : la vie privée et la vie publique des Grecs*, 1894 XXIII
HOMO (Léon), *Problèmes sociaux de jadis et d'à présent*, Paris, 1922 XXIV
MEYER (Eduard), *Geschichte des Altertums*, 3rd ed., Stuttgart, 1913 XXV
BELOCH (Julius), *Griechische Geschichte*, 2nd ed., Strasburg, 1912 XXVI
BURCKHARDT (J.), *Griechische Kulturgeschichte*, Berlin and Stuttgart, 1898 XXVII
BURY (J. B.), *History of Greece to the death of Alexander the Great*, London, 1902 XXVIII
BUSOLT (B.), *Griechische Geschichte*, 2nd ed., Gotha, 1893-1904 XXIX
CURTIUS (Ernst), [*Griechische Geschichte*, 6th ed., Berlin, 1887-89; *The History of Greece*, translated by A. W. WARD, London, 1868-73] *Histoire grecque*, French trans., 1880, etc. XXX
DROYSEN (Johann G.) [*Geschichte des Hellenismus*, Hamburg, 1836] *Histoire de l'Hellénisme*, French trans., Paris, 1883-85 XXXI
HERTZBERG (Gustav F.), [*Die Geschichte Griechenlands unter der Herrschaft der Römer*, Halle, 1866-75] *Histoire de la Grèce sous la domination des Romains*, French trans., Paris, 1880, etc. XXXII
HATZFELD (J.), *Histoire grecque*, 1926 XXXIII
MOMMSEN (Theodor), *Histoire romaine*, French trans., Paris, 1863-89 [*The History of Rome*, and *The Provinces of the Roman Empire*, translated by W. P. DICKSON, London, 1872, etc.; 1909] XXXIV

Bluemner (Hugo), *Die gewerbliche Tätigkeit der Völker des klassischen Alterthums*, Leipzig, 1869 . . XXXV
Dareste (R.), *Nouvelles Études sur l'histoire du droit*, 1902 XXXVI
Fustel de Coulanges (Numa D.), *Recherches sur quelques problèmes d'histoire*, 1885 XXXVII
— *Nouvelles Recherches sur quelques problèmes d'histoire*, 1891 XXXVIII

IV.—SPECIAL WORKS

Book I

Ardaillon (E.), *Les Mines du Laurion*, Paris, 1897 . XXXIX
Babelon (Ernest), *Les Monnaies grecques*, 1921 . . XL
Cavaignac (Eugène), *Études sur l'histoire financière d'Athènes au Ve siècle*, Paris, 1908 . . . XLI
Francotte (H.), *L'Industrie dans la Grèce ancienne*, 1900, etc. XLII
Gernet (L.), *L'Approvisionnement d'Athènes en blé au Ve et au IVe siècle*, Paris, 1909 . . . XLIII
Glotz (Gustave), *Le Travail dans la Grèce ancienne*, Paris, 1920 [*Ancient Greece at Work*, translated by M. R. Dobie (*History of Civilization*), London and N.Y., 1926] XLIV
Guiraud (Paul), *La Propriété foncière en Grèce*, Paris, 1893 XLV
— *La Main-d'œuvre industrielle dans l'ancienne Grèce*, Paris, 1900 XLVI
Jardé (A.), *La Formation du peuple grec*, Paris, 1923 [*The Formation of the Greek People*, translated by M. R. Dobie (*History of Civilization*), London and N.Y., 1926] XLVII
— *Les Céréales dans l'antiquité grecque*, Paris, 1925 . XLVIII
Pottier (Edmond), *Les Statuettes de terre cuite dans l'antiquité*, Paris, 1890 XLIX

Book II

Anatolian Studies, presented to Sir William M. Ramsay, Manchester, 1923 L
Babelon (Ernest), *Inventaire sommaire de la Collection Waddington*, 1898 LI
Beloch (Julius), *Die Bevölkerung der griechisch-römischen Welt*, Leipzig, 1886 LII
Bouché-Leclercq (A.), *Histoire des Lagides*, Paris, 1903-1907 LIII
Chapot (Victor), *La Province romaine proconsulaire d'Asie*, Paris, 1904 LIV
Hartmann (Mlle.), *L'Agriculture dans l'ancienne Égypte*, 1923 LV
Hatzfeld (J.), *Les Trafiquants italiens dans l'Orient hellénique*, 1919 LVI

HAUSSOULLIER (Bernard), *Études sur l'histoire de Milet et du Didymeion*, Paris, 1902 LVII

JOUGUET (Pierre), *L'Impérialisme macédonien et l'hellénisation de l'Orient*, Paris, 1926 [*Macedonian Imperialism*, translated by M. R. DOBIE (*History of Civilization*), London and N.Y., 1928] LVIII

LUMBROSO (G.), *Recherches sur l'économie politique de l'Égypte sous les Lagides*, Turin, 1870 . . . LIX

MAHAFFY (John P.), *A History of Egypt under the Ptolemaic Dynasty*, London, 1899 LX

MASPERO (Henri), *Les Finances de l'Égypte sous les Lagides*, Paris, 1905 LXI

MONTET (P.), *Les Scènes de la vie privée dans les tombeaux égyptiens de l'Ancien Empire*, 1925 . . . LXII

RADET (G.), *De Coloniis a Macedonibus in Asiam cis Taurum deductis*, Paris, 1897 LXIII

RAMSAY (Sir William M.), *The Cities and Bishoprics of Phrygia*, Oxford, 1895-97 LXIV

REIL (T.), *Beiträge zur Kenntniss des Gewerbes im hellenistischen Ægypten*, Leipzig, 1913 LXV

ROSTOVTZEV (Michael), *A Large Estate in Egypt in the third century* B.C., Madison, 1917 . . . LXVI

ROUSSEL (P.), *Délos colonie athénienne*, Paris, 1916 . LXVII

BOOK III

BLOCH (Gustave), *La République romaine*, Paris, 1913 . LXVIII

FRANK (Tenney), *An Economic History of Rome to the End of the Republic*, Baltimore, 1920 [2nd ed., 1927] . . LXIX

GRENIER (Albert), *Bologne villanovienne et étrusque*, Paris, 1914 LXX

GSELL (Stéphane), *Histoire ancienne de l'Afrique du Nord*, Paris, 1913, etc. LXXI

HELBIG (W.), *Die Italiker in der Poebene*, 1879 . . LXXII

HOMO (Léon), *L'Italie primitive et les débuts de l'impérialisme romain*, Paris, 1925 [*Primitive Italy*, translated by V. G. CHILDE (*History of Civilization*), London and N.Y., 1927] LXXIII

MELTZER (O.), *Geschichte der Karthager*, 1879, etc. . LXXIV

MERLIN (A.), *L'Aventin dans l'antiquité*, Paris, 1906 . LXXV

MOMMSEN (Theodor), *Histoire de la monnaie romaine*, French trans., 1865-75 LXXVI

MONTELIUS (O.), *La Civilisation primitive en Italie*, 1895, etc. LXXVII

MORGAN (Jacques de), *L'Humanité préhistorique*, Paris, 1921 [*Prehistoric Man*, translated by J. H. PAXTON and V. C. C. COLLUM (*History of Civilization*), London and N.Y., 1924] LXXVIII

PAIS (E.), *Italia antica*, 1920, [*Ancient Italy*, translated by C. D. CURTIUS, Chicago and London, 1908] . . LXXIX

PEET (T. E.), *The Stone and Bronze Ages in Italy and Sicily*, 1909 LXXX

PIGANIOL (A.), *Essai sur les origines de Rome*, Paris, 1917 . LXXXI

Book IV

Blanchet (Adrien), *Les Trésors de monnaies romaines et les invasions germaniques en Gaule*, Paris, 1900 . . LXXXII

Bloch (Gustave), *L'Empire romain*, Paris, 1922 . . LXXXIII

Cagnat (René), *L'Armée romaine d'Afrique*, 2nd ed., Paris, 1912 LXXXIV

— *Étude historique sur les impôts indirects chez les Romains*, 1882 LXXXV

— *Le Portorium chez les Romains*, 1880 . . . LXXXVI

Chapot (Victor), *Le Monde romain*, Paris, 1927 [*The Roman World*, translated by E. A. Parker (*History of Civilization*), London and N.Y., 1928] . . LXXXVII

Charlesworth (M. P.), *Trade-Routes and Commerce of the Roman Empire*, Cambridge, 1924 . . . LXXXVIII

Collingwood (R. G.), *Roman Britain*, Oxford, 1923 . LXXXIX

Cumont (Franz), *Comment la Belgique fut romanisée*, Brussels, 1919 XC

Cuq (E.), *Institutions juridiques des Romains*, 1917 . XCI

Ferrero (Guglielmo), *La Ruine de la civilisation antique*, Paris, 1921 [*The Ruin of the Ancient Civilisation*, translated by the Hon. Lady Whitehead, N.Y., 1921] . XCII

Frank (Tenney), *Roman Imperialism*, N.Y., 1914 . XCIII

Haverfield (F.), *The Romanization of Roman Britain*, 4th ed., Oxford, 1923 XCIV

Homo (Léon), *L'Empire romain*, Paris, 1925 . . XCV

Jullian (Camille), *Histoire de la Gaule*, Paris, 1908, etc. . XCVI

Peyre (R.), *L'Empire romain*, 1894 . . . XCVII

Rostovtzev (Michael), *The Social and Economic History of the Roman Empire*, Oxford, 1926 . . XCVIII

— *Studien zur Geschichte des römischen Kolonates*, Berlin and Leipzig, 1910 XCIX

— *Geschichte der Staatspacht in der römischen Kaiserzeit bis Diokletian*, 1902 C

Sagot (François), *La Bretagne romaine*, Paris, 1911 . CI

Seeck (Otto), *Geschichte des Untergangs der antiken Welt*, Berlin, 1895-1921 CII

Vaglieri (D.), *Ostia : cenni storici e guida*, Rome, 1914 . CIII

Waltzing (J. P.), *Étude historique sur les corporations professionelles chez les Romains*, Louvain, 1895-1902 . CIV

V.—HISTORICAL ATLASES

Atlas de géographie historique, published under the direction of M. Schrader, 1896 CV

Atlas général Vidal de la Blache, 1894 . . CVI

Spruner Menke, *Atlas Antiquus*, 1862 . . . CVII

Sieglin (W.), *Atlas Antiquus* CVIII

INDEX

Note.—*Greek names are transliterated on the usual English system, Roman terminations being avoided for choice. Romans are indexed under their cognomen, except those usually known by their gentile name. Men indexed under the same name are given in order of date.*

THE HISTORY OF CIVILIZATION

Titles in the series

Pre History	Language - A Linguistic Introduction to History	*J Vendryes*
	A Geographical Introduction to History	*Lucien Febvre*
	The Dawn of European Civilization	*V Gordon Childe*
	The Aryans	*V Gordon Childe*
	From Tribe to Empire	*Moret & Davy*
	Death Customs	*Effie Bendann*
	The Migration of Symbols	*D Mackenzie*
	The History of Witchcraft and Demonology	*Montague Summers*
	The History of Medicine	*C G Cumston*
	Money and Monetary Policy in Early Times	*A R Burns*
	Life and Work in Prehistoric Times	*G Renard*
	Social Organization	*Rivers & Perry*
Greek Civilization	The Ægean Civilization	*G Glotz*
	Ancient Greece at Work	*G Glotz*
	The Formation of the Greek People	*A Jardé*
	Art in Greece	*de Ridder & Deonna*
	Macedonian Imperialism	*Pierre Jouguet*
	Greek Thought and the Origins of the Scientific Spirit	*Léon Robin*
	The Greek City and its Institutions	*G Glotz*
Roman Civilization	Primitive Italy	*Leon Homo*
	Rome the Law-Giver	*J Declareuil*
	The Roman Spirit	*Albert Grenier*
	The Roman World	*V Chapot*
	Roman Political Institutions	*Leon Homo*
	The Economic Life of the Ancient World	*J Toutain*
Eastern Civilization	The Nile and Egyptian Civilization	*A Moret*
	The Peoples of Asia	*L H Dudley Buxton*
	Mesopotamia	*L Delaporte*
	A Thousand Years of the Tartars	*E H Parker*
	Ancient Persia and Iranian Civilization	*Clement Huart*
	Chinese Civilization	*Marcel Granet*
	The Life of Buddha	*Edward J Thomas*
	The History of Buddhist Thought	*Edward J Thomas*
	Ancient India and Indian Civilization	*Masson-Oursel et al*
	The Heroic Age of India	*N K Sidhanta*
Judaeo Christian Civilization		
	Israel	*Adolphe Lods*
	The Prophets and the Rise of Judaism	*Adolphe Lods*
	The Jewish World in the Time of Jesus	*Charles Guignebert*
	The History and Literature of Christianity	*Pierre de Labriolle*
European Civilization	The End of the Ancient World	*Ferdinand Lot*
	The Rise of the Celts	*Henri Hubert*
	The Greatness and Decline of the Celts	*Henri Hubert*
	Life and Work in Medieval Europe	*P Boissonnade*
	The Feudal Monarchy in France and England	*C Petit-Dutaillis*
	Travel and Travellers of the Middle Ages	*Arthur Newton*
	Chivalry	*Edgar Prestage*
	The Court of Burgundy	*Otto Cartellieri*
	Life and Work in Modern Europe	*Renard & Weulersse*
	China and Europe	*Adolf Reichwein*
	The American Indian Frontier	*W Christie Macleod*